# Get the eBook FREE!

(PDF, ePub, Kindle, and liveBook all included)

We believe that once you buy a book from us, you should be able to read it in any format we have available. To get electronic versions of this book at no additional cost to you, purchase and then register this book at the Manning website.

Go to https://www.manning.com/freebook and follow the instructions to complete your pBook registration.

## That's it!
## Thanks from Manning!

*Event Streams in Action*

# Event Streams in Action

### REAL-TIME EVENT SYSTEMS WITH KAFKA AND KINESIS

ALEXANDER DEAN
VALENTIN CRETTAZ

MANNING

SHELTER ISLAND

Manning Publications Co.
20 Baldwin Road
PO Box 761
Shelter Island, NY 11964

| | |
|---|---|
| Acquisitions editors: | Mike Stephens and Frank Pohlmann |
| Development editors: | Jennifer Stout and Cynthia Kane |
| Technical development editor: | Kostas Passadis |
| Review editor: | Aleks Dragosacljević |
| Production editor: | Anthony Calcara |
| Copy editor: | Sharon Wilkey |
| Proofreader: | Melody Dolab |
| Technical proofreader: | Michiel Trimpe |
| Typesetter: | Dennis Dalinnik |
| Cover designer: | Marija Tudor |

ISBN: 9781617292347
Printed and bound by CPI Group (UK) Ltd, Croydon, CR0 4YY

# brief contents

# contents

vii

# *preface*

A continuous stream of real-world and digital events already power the company where you work, even though you probably don't think in those terms. Instead, you likely think about your daily work in terms of the people or things that you interact with, the software or hardware you use to get stuff done, or your own microcosm of a to-do list of tasks.

Computers can't think like this! Instead, computers see a company as an organization that generates a response to a continuous stream of events. We believe that reframing your business in terms of a continuous stream of events offers huge benefits. This is a young but hugely important field, and there is a lot still to discuss.

*Event Streams in Action* is all about events: how to define events, how to send streams of events into unified log technologies like Apache Kafka and Amazon Kinesis, and how to write applications that process those event streams. We're going to cover a lot of ground in this book: Kafka and Kinesis, stream processing frameworks like Samza and Spark Streaming, event-friendly databases like Amazon Redshift, and more.

This book will give you confidence to identify, model, and process event streams wherever you find them—and we guarantee that by the end of this book, you will be seeing event streams everywhere! Above all, we hope that this book acts as a springboard for a broader conversation about how we, as software engineers, should work with events.

# acknowledgments

I would like to thank my wife Charis for her support through the long process of writing this book, as well as my parents for their lifelong encouragement. And many thanks to my cofounder at Snowplow Analytics, Yali Sassoon, for giving me the "air cover" to work on this book even while we were trying to get our tech startup off the ground.

On the Manning side, I will always be appreciative to commissioning editor Frank Pohlmann for believing I had a book in me. Thanks too to Cynthia Kane, Jennifer Stout, and Rebecca Rinehart for their patience and support through the difficult and lengthy gestation. I am grateful to my coauthor, Valentin Crettaz, for his contributions and his laser focus on getting this book completed. Special thanks also to all the reviewers whose feedback and insight greatly helped to improve this book, including Alex Nelson, Alexander Myltsev, Azatar Solowiej, Bachir Chihani, Charles Chan, Chris Snow, Cosimo Attanasi, Earl Bingham, Ernesto Garcia, Gerd Klevesaat, Jeff Lim, Jerry Tan, Lourens Steyn, Miguel Eduardo Gil Biraud, Nat Luengnaruemitchai, Odysseas Pentakalos, Rodrigo Abreu, Roger Meli, Sanket Naik, Shobha Iyer, Sumit Pal, Thomas Lockney, Thorsten Weber, Tischliar Ronald, Tomasz Borek, and Vitaly Bragilevsky.

Finally, I'd like to thank Jay Kreps, CEO of Confluent and creator of Apache Kafka, for his monograph "The Log," published back in December 2013, which started me on the journey of writing this book in addition to informing so much of my work at Snowplow.

—ALEXANDER DEAN

First and foremost, I'd like to thank my family for having to deal daily with a father and husband who is so passionate about his work that he sometimes (read: often) forgets to give his keyboards and mice a break. I would never have been able to fulfill my dreams without your unconditional support and understanding.

I've worked with Manning on many different book projects over a long period of time now. But this one was special—not only a nice technological adventure, but also a human one. I can't emphasis the human part enough, as writing books is not only about content, grammar rules, typos, and phrasing, but also about collaborating and empathizing with human beings, understanding their context and their sensibilities, and sharing one chapter of your life with them. For all this, I'd like to thank Michael Stephens, Jennifer Stout, and Rebecca Rinehart for taking the time and effort to persuade me to take on this project. It wasn't easy (it never is and never should be), but it was a great deal of fun and highly instructive.

Finally, I'd like to thank Alex for being such a good writer and for always managing to mix an entertaining writing style with illustrative examples and figures to make complex subjects and concepts easy for the reader to grasp.

—VALENTIN CRETTAZ

# *about this book*

Writing real-world applications in a data-rich environment can feel like being caught in the cross fire of a paintball battle. Any action may require you to combine event streams, batch archives, and live user or system requests in real time. Unified log processing is a coherent data processing architecture designed to encompass batch and near-real-time stream data, event logging and aggregation, and data processing on the resulting unified event stream. By efficiently creating a single log of events from multiple data sources, unified log processing makes it possible to design large-scale data-driven applications that are easier to design, deploy, and maintain.

## Who should read this book

This book is written for readers who have experience writing some Java code. Scala and Python experience may be helpful to understanding some concepts in the book but is not required.

## How this book is organized: a roadmap

This book has 11 chapters divided into three parts.

Part 1 defines event streams and unified logs, providing a wide-ranging look:

- Chapter 1 provides a ground-level foundation by offering definitions and examples of events and continuous event streams, and takes a brief look at unifying event streams with a unified log.
- Chapter 2 dives deep into the key attributes of a unified log, and walks you through setting up, sending, and reading events in Apache Kafka.

- Chapter 3 introduces event stream processing, and how to write applications that process individual events while also validating and enriching events.
- Chapter 4 focuses on event stream processing with Amazon Kinesis, a fully managed unified log service.
- Chapter 5 looks at stateful stream processing, using the most popular stream processing frameworks to process multiple events from a stream-using state.

Part 2 dives deep into the quality of events being fed into a unified log:

- Chapter 6 covers event schemas and schema technologies, focusing on using Apache Avro to represent self-describing events.
- Chapter 7 covers event archiving, providing a deep look into why archiving a unified log is so important and the best practices for doing so.
- Chapter 8 looks at how to handle failure in Unix programs, Java exceptions, and error logging, and how to design for failure inside and across stream processing applications.
- Chapter 9 covers the role of commands in the unified log, using Apache Avro to define schemas and process commands.

Part 3 takes an analysis-first look at the unified log, leading with the two main methodologies for unified log analytics, and then applying various database and stream processing technologies to analyze our event streams:

- Chapter 10 uses Amazon Redshift, a horizontally scalable columnar database, to cover analytics-on-read versus analytics-on-write and techniques for storing and widening events.
- Chapter 11 provides simple algorithms for analytics-on-write event streams, and will allow you to deploy and test an AWS Lambda function.

## About the code

This book contains many examples of source code, both in numbered listings and in line with normal text. In both cases, source code is formatted in a `fixed-width font like this` to separate it from ordinary text. Sometimes code is also **in bold** to highlight code that has changed from previous steps in the chapter, such as when a new feature adds to an existing line of code.

In many cases, the original source code has been reformatted; we've added line breaks and reworked indentation to accommodate the available page space in the book. In rare cases, even this was not enough, and listings include line-continuation markers (➥). Additionally, comments in the source code have often been removed from the listings when the code is described in the text. Code annotations accompany many of the listings, highlighting important concepts.

Source code for the examples in this book is available for download from the publisher's website at www.manning.com/books/event-streams-in-action.

### liveBook discussion forum

Purchase of *Event Streams in Action* includes free access to a private web forum run by Manning Publications, where you can make comments about the book, ask technical questions, and receive help from the authors and from other users. To access the forum, go to https://livebook.manning.com/#!/book/event-streams-in-action/discussion. You can also learn more about Manning's forums and the rules of conduct at https://livebook.manning.com/#!/discussion.

Manning's commitment to our readers is to provide a venue where a meaningful dialogue between individual readers and between readers and the author can take place. It is not a commitment to any specific amount of participation on the part of the author, whose contribution to the forum remains voluntary (and unpaid). We suggest you try asking the author some challenging questions lest his interest stray! The forum and the archives of previous discussions will be accessible from the publisher's website as long as the book is in print.

# *about the authors*

ALEXANDER DEAN is cofounder and technical lead of Snowplow Analytics, an open source event processing and analytics platform.

VALENTIN CRETTAZ is an independent IT consultant who's been working for the past 25 years on many challenging projects across the globe. His expertise ranges from software engineering and architecture to data science and business intelligence. His daily job boils down to using the latest and most cutting-edge web, data, and streaming technologies to implement IT solutions that will help reduce the cultural gap between IT and business people.

# *about the cover illustration*

The figure on the cover of *Event Streams in Action* is captioned "Habit of a Lady of Tartary in 1667." The illustration is taken from Thomas Jefferys' *A Collection of the Dresses of Different Nations, Ancient and Modern* (four volumes), London, published between 1757 and 1772. The title page states that these are hand-colored copperplate engravings, heightened with gum arabic.

Thomas Jefferys (1719–1771) was called "Geographer to King George III." He was an English cartographer who was the leading map supplier of his day. He engraved and printed maps for government and other official bodies and produced a wide range of commercial maps and atlases, especially of North America. His work as a map maker sparked an interest in local dress customs of the lands he surveyed and mapped, which are brilliantly displayed in this collection. Fascination with faraway lands and travel for pleasure were relatively new phenomena in the late eighteenth century, and collections such as this one were popular, introducing both the tourist as well as the armchair traveler to the inhabitants of other countries.

The diversity of the drawings in Jefferys' volumes speaks vividly of the uniqueness and individuality of the world's nations some 200 years ago. Dress codes have changed since then, and the diversity by region and country, so rich at the time, has faded away. It's now often hard to tell the inhabitants of one continent from another. Perhaps, trying to view it optimistically, we've traded a cultural and visual diversity for a more varied personal life—or a more varied and interesting intellectual and technical life.

At a time when it's difficult to tell one computer book from another, Manning celebrates the inventiveness and initiative of the computer business with book covers based on the rich diversity of regional life of two centuries ago, brought back to life by Jefferys' pictures.

# Part 1

# Event streams and unified logs

In this first part, we'll introduce the basics of event streaming and explain what a unified log is. We'll also show how to use technologies such as Apache Kafka, Amazon Kinesis, and Apache Samza in order to process event streams.

# Introducing event streams

**This chapter covers**

- Defining events and continuous event streams
- Exploring familiar event streams
- Unifying event streams with a unified log
- Introducing use cases for a unified log

Believe it or not, a continuous stream of real-world and digital events already powers the company where you work. But it's unlikely that many of your coworkers think in those terms. Instead, they probably think about their work in terms of the following:

- The people or things that they interact with on a daily basis—for example, customers, the Marketing team, code commits, or new product releases
- The software and hardware that they use to get stuff done
- Their own daily inbox of tasks to accomplish

People think and work in these terms because people are not computers. It is easy to get up in the morning and come to work because Sue in QA really needs those reports for her boss by lunchtime. If we stopped and started to think about our work as creating and responding to a continuous stream of events, we would probably go a little crazy—or at least call in to the office for a duvet day.

3

Computers don't have this problem. They would be comfortable with this definition of a business:

*A company is an organization that generates and responds to a continuous stream of events.*

This definition is not going to win any awards from economists, but we, the authors, believe that reframing your business in terms of a continuous stream of events offers huge benefits. Specifically, event streams enable the following:

- *Fresher insights*—A continuous stream of events represents the "pulse" of a business and makes a conventional batch-loaded data warehouse look stale in comparison.
- *A single version of the truth*—Ask several coworkers the same question and you may well get different answers, because they are working from different "pots" of data. Well-modeled event streams replace this confusion with a single version of the truth.
- *Faster reactions*—Automated near-real-time processing of continuous event streams allows a business to respond to those events within minutes or even seconds.
- *Simpler architectures*—Most businesses have built up a bird's nest of bespoke point-to-point connections between their various transactional systems. Event streams can help to unravel these messy architectures.

Some of these benefits may not seem obvious now, but don't worry: in this chapter, we will go back to first principles, starting with what we mean by *events*. We will introduce some simple examples of events, and then explain what a *continuous event stream* really is. There's a good chance you will find that you are pretty comfortable working with event streams already—you just haven't thought of them in those terms.

Once we have presented some familiar event streams, we will zoom out a level and explain how businesses' handling of events has evolved over the past 20 years. You will see that successive waves of technology have made things much more complex than they should be, but that a new architectural pattern called the *unified log* promises to simplify things again.

For these new approaches to reach the mainstream, they must be backed up with compelling use cases. We will make the benefits of continuous event streams and the unified log significantly more real with a set of tangible real-world use cases, across a variety of industries.

## 1.1   *Defining our terms*

If you work in any kind of modern-day business, chances are that you have already worked with event streams in various forms but have not been introduced to them as such. This section presents a simple definition for an event and then explains how events combine into a continuous event stream.

### 1.1.1 Events

Before we can define a continuous event stream, we need to break out of Synonym City and concretely define a single event. Fortunately, the definition is simple: an *event* is anything that we can observe occurring at a particular point in time. That's it, *fin.* Figure 1.1 sets out four example events from four different business sectors.

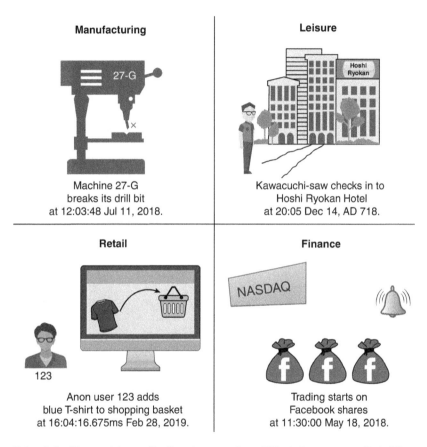

**Manufacturing**

Machine 27-G
breaks its drill bit
at 12:03:48 Jul 11, 2018.

**Leisure**

Kawacuchi-saw checks in to
Hoshi Ryokan Hotel
at 20:05 Dec 14, AD 718.

**Retail**

Anon user 123 adds
blue T-shirt to shopping basket
at 16:04:16.675ms Feb 28, 2019.

**Finance**

Trading starts on
Facebook shares
at 11:30:00 May 18, 2018.

**Figure 1.1  The precision on the timestamps varies a little, but you can see that all four of these events are discrete, recordable occurrences that take place in the physical or digital worlds (or both).**

It is easy to get carried away with the simplicity of this definition of an event, so before we go any further, let's clarify what is not an event. This is by no means an exhaustive list, but these are some of the more common mistakes to avoid. An event is not any of the following:

- *A description of the ongoing state of something*—The day was warm; the car was black; the API client was broken. But "the API client broke at noon on Tuesday" is an event.

- *A recurring occurrence*—The NASDAQ opened at 09:30 every day in 2018. But each individual opening of the NASDAQ in 2018 is an event.
- *A collection of individual events*—The Franco-Prussian war involved the Battle of Spicheren, the Siege of Metz, and the Battle of Sedan. But "war was declared between France and Prussia on 19 July 1870" is an event.
- *A happening that spans a time frame*—The 2018 Black Friday sale ran from 00:00:00 to 23:59:59 on November 23, 2018. But the beginning of the sale and the end of the sale are events.

Here's a general rule of thumb: if the thing you are describing can be tied to a specific point in time, chances are that you are describing an event of some kind, even if it needs some verbal gymnastics to represent it.

## 1.1.2 Continuous event streams

Now that we have defined what an event is, what is a continuous event stream? Simply put, a *continuous event stream* is an unterminated succession of individual events, ordered by the point in time at which each event occurred. Figure 1.2 sketches out what a continuous event stream looks like at a high level: you can see a succession of individual events, stepping forward in time.

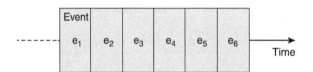

**Figure 1.2   Anatomy of a continuous event stream: time is progressing left to right, and individual events are ordered within this time frame. Note that the event stream is unterminated; it can extend in both directions beyond our ability to process it.**

We say that the succession of events is *unterminated,* because of these facts:

- The start of the stream may predate our observing of the stream.
- The end of the stream is at some unknown point in the future.

To illustrate this, let's consider guests checking into the Hoshi Ryokan hotel in Japan. Hoshi Ryokan is one of the oldest businesses in the world, having been founded in AD 718. Whatever stream of guest check-in events we could analyze for Hoshi Ryokan, we would know that the oldest guest check-ins are lost in the mists of time, and that future check-in events will continue to occur long after we have retired.

## 1.2 Exploring familiar event streams

If you read the previous section and thought that events and continuous event streams seemed familiar, then chances are that you have already worked with event streams, although they were likely not labeled as such. A huge number of software systems are heavily influenced by the idea of generating and responding to a continuous stream of events, including these:

- *Transactional systems*—Many of these respond to external events, such as customers placing orders or suppliers delivering parts.
- *Data warehouses*—These collect the event histories of other systems for later analysis, storing them in *fact tables*.
- *Systems monitoring*—This continually checks system- and application-level events coming from software or hardware systems to detect issues.
- *Web analytics packages*—Through these, analysts can explore website visitors' on-site event streams to generate insights.

In this section, we will take a brief tour through three common areas of programming in which the event stream concept is close to the surface. Hopefully, this will make you think about part of your existing toolkit in a more event-centric way. But if all of these examples are unfamiliar to you, don't worry: you'll have plenty of opportunities to master event streams from first principles later.

### 1.2.1 Application-level logging

Let's start with the event stream that almost all backend (and many frontend) developers will be familiar with: application-level logging. If you have worked with Java, chances are that you have worked with Apache Log4j at one time or another, but if not, don't worry: its approach to logging is pretty similar to lots of other tools. Assuming that the Log4j.properties file is correctly configured and a static logger is initialized, logging with Log4j is simple. The following listing sets out examples of log messages that a Java developer might add to their application.

**Listing 1.1   Application logging with Log4j**

```
doSomethingInteresting();
log.info("Did something interesting");
doSomethingLessInteresting();
log.debug("Did something less interesting");

// Log output:                              <———————————————┐
// INFO  2018-10-15 10:50:14,125 [Log4jExample_main]        |
  "org.alexanderdean.Log4jExample": Did something interesting |
// INFO  2018-10-15 10:55:34,345 [Log4jExample_main]        |
  "org.alexanderdean.Log4jExample": Did something less interesting
```

**The log output format assumes that we configured our Log4j.properties file like so: log4j.appender.stdout.layout.ConversionPattern=%-5p %d [%t] %c: %m%n.**

You can see that application-level logging is generally used to record specific events at a point in time. The log events expressed in the code are deliberately primitive, consisting of just a log level indicating the severity of the event, and a message string describing the event. But Log4j does add metadata behind the scenes; in this case, the time of the event and the reporting thread and class name.

What happens to the log events after they are generated by your application? Best practice says that you write the log events to disk as log files, and then use a log collection technology, such as Flume, Fluentd, Logstash, or Filebeat, to collect the log files from the individual servers and ingest them into a tool for systems monitoring or log-file analysis. Figure 1.3 illustrates this event stream.

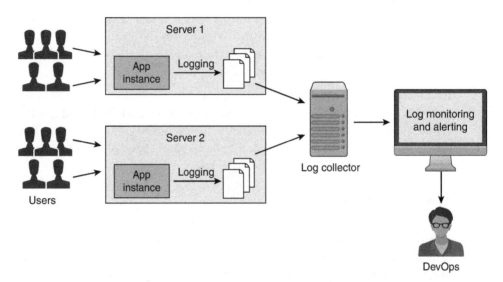

**Figure 1.3   An application is running on two servers, with each application instance generating log messages. The log messages are written (*rotated*) to disk before being collected and forwarded to a systems-monitoring or log-file-analysis tool.**

So application-level logging is clearly a continuous event stream, albeit one that leans heavily on schemaless messages that are often only human-readable. As the Log4j example hints at, application-level logging is highly configurable, and not well standardized across languages and frameworks. When working on a polyglot project, standardizing with a common log format across all software can be painful.

### 1.2.2   Web analytics

Let's move on to another example. If you are a frontend web developer, there's a good chance that you have embedded JavaScript tags in a website or web app to provide some kind of web or event analytics. The most popular software in this category is Google Analytics, a software-as-a-service (SaaS) web analytics platform from Google; in 2012, Google released a new iteration of its analytics offering called Universal Analytics.

Listing 1.2 shows example JavaScript code used to instrument Universal Analytics. This code would be either embedded directly in the source code of the website or invoked through a JavaScript tag manager. Either way, this code will run for each visitor to the website, generating a continuous stream of events representing each visitor's set of interactions with the website. These events flow back to Google, where they are stored, processed, and displayed in a variety of reports. Figure 1.4 demonstrates the overall event flow.

**Listing 1.2   Web tracking with Universal Analytics**

Initialization code for the Universal Analytics tracking tag

Track the website visitor viewing this web page.

```
<script>
(function(i,s,o,g,r,a,m){i['GoogleAnalyticsObject']=r;i[r]=i[r]||function(){
    (i[r].q=i[r].q||[]).push(arguments)},i[r].l=1*new
        Date();a=s.createElement(o),
    m=s.getElementsByTagName(o)[0];a.async=1;a.src=g;m.parentNode
        .insertBefore(a,m)
}) (window,document,'script','//www.google-analytics.com/analytics.js','ga');

    ga('create', 'UA-34290195-2', 'test.com');
    ga('send', 'pageview');
    ga('send', 'event', 'video', 'play', 'doge-video-01');

</script>
```

Create an event tracker for the given account, for the test.com website.

Track the website visitor watching a video on this web page.

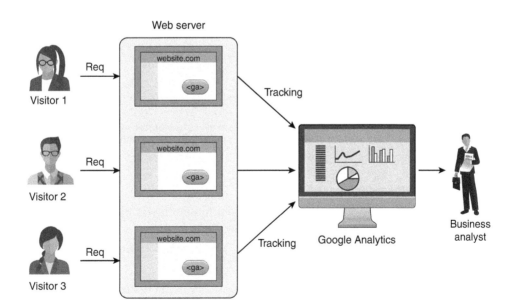

Figure 1.4   A JavaScript tracking tag sends visitors' interactions with a website to Universal Analytics. This event stream is made available for analysis from within the Google Analytics user interface.

With Google Analytics deployed like this, a business analyst can log in to the Google Analytics web interface and start to make sense of the website's event stream across all of its visitors. Figure 1.5 is a screenshot taken from Universal Analytics' real-time dashboard, showing the previous 30 minutes' worth of events occurring on the Snowplow Analytics website.

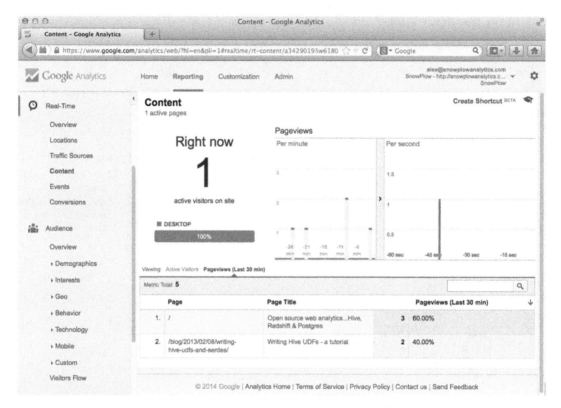

**Figure 1.5   Google Analytics is recording a real-time stream of events generated by website visitors. At the bottom right, you can see the counts of views of individual web pages in the last 30 minutes.**

### 1.2.3   *Publish/subscribe messaging*

Let's take a slightly lower-level example, but hopefully still one that many readers will be familiar with: application messaging, specifically in the publish/subscribe pattern. *Publish/subscribe*, sometimes shortened to *pub/sub*, is a simple way of communicating messages:

- Message senders publish messages that can be associated with one or more topics.
- Message receivers subscribe to specific topics, and then receive all messages associated with that topic.

If you have worked with pub/sub messaging, there's a good chance that the messages you were sending were *events* of some form or another.

For a hands-on example, let's try out NSQ, a popular distributed pub/sub messaging platform originally created by Bitly. Figure 1.6 illustrates NSQ brokering events between a single *publishing* app and two *subscribing* apps.

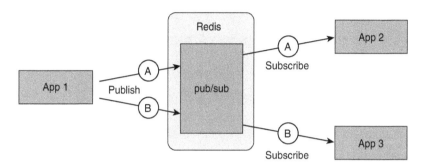

**Figure 1.6  NSQ pub/sub is facilitating communication between App 1, which is publishing messages into a single Topic 1, and Apps 2 and 3, which are each subscribing to receive messages from Topic 1.**

The nice thing about NSQ for demonstration purposes is that it is super simple to install and set up. On macOS, we open up a new terminal, install NSQ by using Homebrew, and then start up the `nsqlookupd` daemon:

```
$ brew install nsq
...
$ nsqlookupd
...
```

And then in a second terminal window, we start the main NSQ daemon, `nsqd`:

```
$ nsqd --lookupd-tcp-address=127.0.0.1:4160
...
```

We leave those two daemons running and then open a third terminal. We use the `nsqd` daemon's HTTP API to create a new topic:

```
$ curl -X POST http://127.0.0.1:4151/topic/create\?topic\=Topic1
```

Next we're ready to create the two subscribers, Apps 2 and 3. In two further terminals, we start the `nswq_tail` app to simulate Apps 2 and 3 subscribing to Topic 1:

```
$ nsq_tail --lookupd-http-address=127.0.0.1:4161 \
  --topic=Topic1 --channel=App2
2018/10/15 20:53:10 INF    1 [Topic1/App2]
 querying nsqlookupd http://127.0.0.1:4161/lookup?topic=Topic1
2018/10/15 20:53:10 INF    1 [Topic1/App2]
 (Alexanders-MacBook-Pro.local:4150) connecting to nsqd
```

And our fifth and final terminal:

```
$ nsq_tail --lookupd-http-address=127.0.0.1:4161 \
  --topic=Topic1 --channel=App3
2018/10/15 20:57:55 INF    1 [Topic1/App3]
 querying nsqlookupd http://127.0.0.1:4161/lookup?topic=Topic1
2018/10/15 20:57:55 INF    1 [Topic1/App3]
 (Alexanders-MacBook-Pro.local:4150) connecting to nsqd
```

Returning to our third terminal (the only one not running a daemon), we send in some events, again using the HTTP API:

```
$ curl -d 'checkout' 'http://127.0.0.1:4151/pub?topic=Topic1'
OK%
$ curl -d 'ad_click' 'http://127.0.0.1:4151/pub?topic=Topic1'
OK%
$ curl -d 'save_game' 'http://127.0.0.1:4151/pub?topic=Topic1'
OK%
```

We check back in our tailing terminals to see the events arriving:

```
2018/10/15 20:59:06 INF    1 [Topic1/App2] querying nsqlookupd
http://127.0.0.1:4161/lookup?topic=Topic1
checkout
ad_click
save_game
```

And the same for App 3:

```
2018/10/15 20:59:08 INF    1 [Topic1/App3] querying nsqlookupd
http://127.0.0.1:4161/lookup?topic=Topic1
checkout
ad_click
save_game
```

So in this pub/sub architecture, we have events being published by one application and being subscribed to by two other applications. Add more events, and again you have a continuous event stream being processed.

Hopefully, the examples in this section have shown you that the concept of the event stream is a familiar one, underpinning disparate systems and approaches including application logging, web analytics, and publish/subscribe messaging. The terminology may be different, but in all three examples, you can see the same building blocks: a structure or schema of events (even if extremely minimal); a way of generating these events; and a way of collecting and subsequently processing these events.

## 1.3   *Unifying continuous event streams*

So far in this chapter, we have introduced the idea of event streams, defined our terms, and highlighted familiar technologies that use event streams in one form or another. This usage is a good start, but hopefully you can see that these technologies are highly fragmented: their evented nature is poorly understood, their event schemas are unstandardized, and their use cases are trapped in separate silos. This section introduces a much more radical—and powerful—approach to using continuous event streams for your business.

Simply put, the argument of this book is that every digital business should be restructured around a process that does the following:

- Collects events from disparate source systems
- Stores them in a *unified log*
- Enables data processing applications to operate on these event streams

This is a bold statement—and one that sounds like a lot of work to implement! What evidence do we have that this is a practical and useful course of action for a business?

This section maps out the historical and ongoing evolution of business data processing, extending up to continuous event streams and this unified log. We have split this evolution into two distinct eras that we have both lived through and experienced firsthand, plus a third era that is soon approaching:

- *The classic era*—The pre-big data, pre-SaaS era of operational systems and batch-loaded data warehouses
- *The hybrid era*—Today's hodgepodge of different systems and approaches
- *The unified era*—An emerging architecture, enabled by processing continuous event streams in a unified log

### 1.3.1 The classic era

In the *classic era*, businesses primarily operated a disparate set of on-premises transactional systems, feeding into a data warehouse; figure 1.7 illustrates this architecture. Each transactional system would feature the following:

- An internal *local loop* for near-real-time data processing
- Its own data silo
- Where necessary, point-to-point connections to peer systems (for example, via APIs or feed import/exports)

A data warehouse would be added to give the management team a much-needed view across these transactional systems. This data warehouse would typically be fed from the transactional systems overnight by a set of batch extract, transform, load (ETL) processes. This data warehouse provided the business with a single version of the truth, with full data history and wide data coverage. Internally, it was often constructed following the star schema style of fact and dimension tables, as popularized by Ralph Kimball.[1]

---

[1] See "Fact Tables and Dimension Tables" by Ralph Kimball (www.kimballgroup.com/2003/01/fact-tables-and-dimension-tables/) for more information about these dimensional modeling techniques.

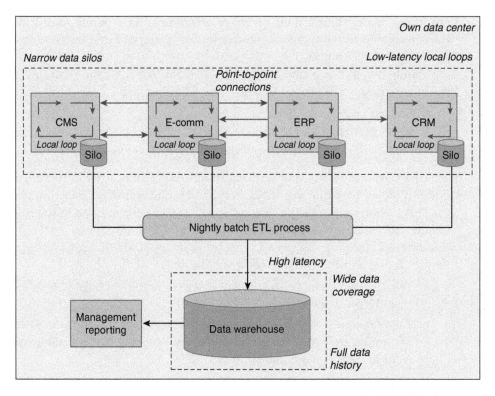

**Figure 1.7    This retailer has four transactional systems, each with its own data silo. These systems are connected to each other as necessary with point-to-point connections. A nightly batch ETL process *extracts* data out of the data silos, *transforms* it for reporting purposes, and then *loads* it into a data warehouse. Management reports are then based on the contents of the data warehouse.**

Although we call this the classic era, in truth many businesses still run on a close descendant of this approach, albeit with more SaaS platforms mixed in. This is a tried and tested architecture, although one with serious pain points:

- *High latency for reporting*—The time span between an event occurring and that event appearing in management reporting is counted in hours (potentially even days), not seconds
- *Point-to-point spaghetti*—Extra transactional systems mean even more point-to-point connections, as illustrated in figure 1.8. This point-to-point spaghetti is expensive to build and maintain and increases the overall fragility of the system.
- *Schema woes*—Classic data warehousing assumes that each business has an intrinsic data model that can be mined from the state stored in its transactional systems. This is a highly flawed assumption, as we explore in chapter 5.

Faced with these issues, businesses have made the leap to a new model—particularly, businesses in fast-moving sectors such as retail, technology, and media. We'll call this new model the hybrid era.

**Maximum point-to-point connections with...**

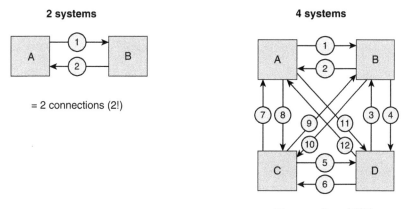

**2 systems**

= 2 connections (2!)

**4 systems**

= 12 connections (4!/2!)

**16 systems**

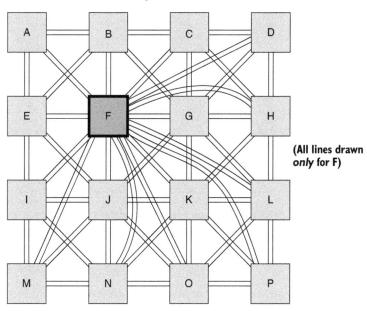

**(All lines drawn *only* for F)**

= 240 connections (16!/14!) (Ouch!)

**Figure 1.8** The maximum number of point-to-point connections possibly required between 2, 4, and 16 software systems is 2, 12, and 240 connections, respectively. Adding systems grows the number of point-to-point connections quadratically.

### 1.3.2  *The hybrid era*

The *hybrid era* is characterized by companies operating a hodgepodge of transactional and analytics systems—some on-premises packages, some from SaaS vendors, plus some homegrown systems. See figure 1.9 for an example of a hybrid-era architecture.

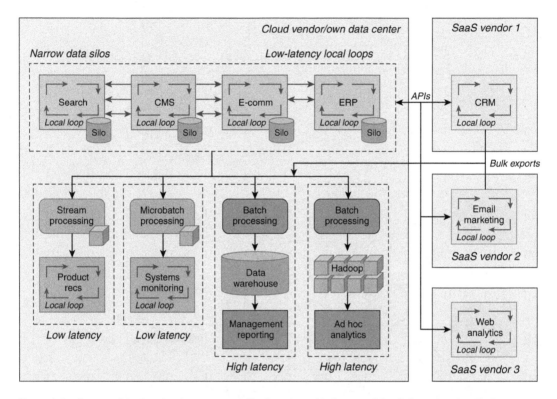

**Figure 1.9  Compared to the classic era, our retailer has now added external SaaS dependencies; Hadoop as a new high-latency, "log everything" platform; and new low-latency data pipelines for use cases such as systems monitoring and product recommendations.**

It is hard to generalize what these hybrid architectures look like. Again, they have strong local loops and data silos, but there are also attempts at "log everything" approaches with Hadoop and/or systems monitoring. There tends to be a mix of near-real-time processing for narrow analytics use cases such as product recommendations, plus separate batch-processing efforts into Hadoop as well as a classic data warehouse. Hybrid architectures also feature attempts to bulk-export data from external SaaS vendors for warehousing, and efforts to feed these external systems with proprietary data through these systems' own APIs.

Although it is obvious that this hybrid approach delivers capabilities sorely lacking from the classic approach, it brings its own problems:

- *No single version of the truth*—Data is now warehoused in multiple places, depending on the data volumes and the analytics latency required. There is no system that has 100% visibility.

- *Decisioning has become fragmented*—The number of local systems loops, each operating on siloed data, has grown since the classic era. These loops represent a highly fragmented approach to making near-real-time decisions from data.

- *Point-to-point connections have proliferated*—As the number of systems has grown, the number of point-to-point connections has exploded. Many of these connections are fragile or incomplete; getting sufficiently granular and timely data out of external SaaS systems is particularly challenging.

- *Analytics can have low latency or wide data coverage, but not both*—When stream processing is selected for low latency, it becomes effectively another local processing loop. The warehouses aim for much wider data coverage, but at the cost of duplication of data and high latency.

### 1.3.3 The unified era

These two eras bring us up to the present day, and the emerging *unified era* of data processing. The key innovation in business terms is putting a unified log at the heart of all of our data collection and processing. A *unified log* is an append-only log to which we write all events generated by our applications. Going further, the unified log has these characteristics:

- Can be read from at low latency.

- Is readable by multiple applications simultaneously, with different applications able to consume from the log at their own pace.

- Holds only a rolling window of events—probably a week or a month's worth. But we can archive the historic log data in the Hadoop Distributed File System (HDFS) or Amazon Simple Storage Service (S3).

For now, don't worry about the mechanics of the unified log. Chapter 2 covers this in much more detail. For now, it is more important to understand how the unified log can reshape the way that data flows through a business. Figure 1.10 updates our retailer's architecture to the unified era. The new architecture is guided by two simple rules:

- All software systems can and should write their individual continuous event streams to the unified log. Even third-party SaaS vendors can emit events via webhooks and streaming APIs.

- Unless very low-latency or transactional guarantees are required, software systems should communicate with each other in an uncoupled way through the unified log, not via point-to-point connections.

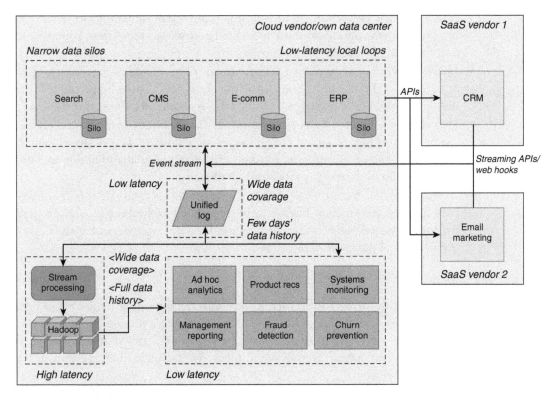

**Figure 1.10   Our retailer has rearchitected around a unified log and a longer-term archive of events in Hadoop. The data architecture is now much simpler, with far fewer point-to-point connections, and all of our analytics and decision-making systems now working off a single version of the truth.**

A few advantages should be clear compared to one or both of the previous architectures:

- *We have a single version of the truth.* Together, the unified log plus Hadoop archive represent our single version of the truth. They contain exactly the same data—our event stream—but they have different time windows of data.

- *The single version of the truth is upstream from the data warehouse.* In the classic era, the data warehouse provided the single version of the truth, making all reports generated from it consistent. In the unified era, the log provides the single version of the truth; as a result, operational systems (for example, recommendation and ad-targeting systems) compute on the same truth as analysts producing management reports.

- *Point-to-point connections have largely been unravelled.* In their place, applications can append to the unified log, and other applications can read their writes. This is illustrated in figure 1.11.

- *Local loops have been unbundled.* In place of local silos, applications can collaborate on near-real-time decision-making via the unified log.

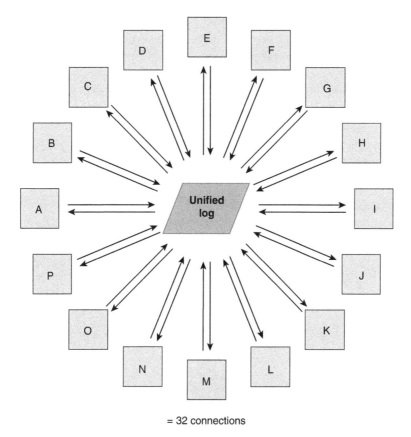

= 32 connections

**Figure 1.11  Our unified log acts as the glue between all of our software systems. In place of the proliferation of point-to-point connections seen prior, we now have systems reading and writing to the unified log. Conceptually, we now have a maximum of 32 unidirectional connections, compared to 240 for a point-to-point approach.**

## 1.4    Introducing use cases for the unified log

You may have read through the preceding section and thought, "Continuous event streams and the unified log look all well and good, but they seem like an architectural optimization, not something that enables wholly new applications." In fact, this is both a significant architectural improvement on previous approaches, and an enabler for powerful new use cases. This section will whet your appetite with three of these use cases.

### 1.4.1    Customer feedback loops

One of the most exciting use cases of continuous data processing is the ability to respond to an individual's customer behavior while that customer is *still engaged with your service.* These real-time feedback loops will look a little different depending on the industry you are in. Here are just a few examples:

- *Retail*—Whenever the customer looks like they are about to abandon their shopping cart, pop up a coupon in their web browser to coax them into checking out. Figure 1.12 shows an example.
- *TV*—Adjust the electronic program guide in real time based on the viewer's current behavior and historical watching patterns, to maximize their viewing hours.
- *Automotive*—Detect abnormal driving patterns and notify the owner that the car may have been stolen.
- *Gaming*—If a player is finding a four-player cooperative game too challenging, adjust the difficulty level to prevent them from quitting and spoiling the game for the other players.

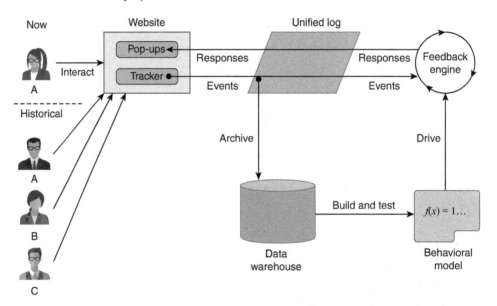

**Figure 1.12   We can use a unified log to respond to customers in an e-commerce environment. The event stream of customer behavior is fed to the unified log and onward to the data warehouse for analysis. A behavioral model is constructed using the historic data, and then this is used to drive a feedback engine. The feedback engine communicates back to the site again via the unified log.**

Customer feedback loops are not new; even in the classic era, you could capture user behavior and use that to inform a personalized mail-out or email drip marketing. And there are startups today that will put a JavaScript tag in your website, track a user in real-time, and attempt to alter their behavior with banners, flash messages, and pop-ups. But the feedback loops enabled by a unified log are much more powerful:

- They are fully under the control of the service provider, not a third party such as a SaaS analytics provider, meaning that these loops can be programmed to whatever algorithm makes the most sense to the business itself.
- They are driven off models tested across the full archive of the exact same event stream.

- Customers' reactions to the feedback loops can be added into the event stream as well, making machine-learning approaches possible.

### 1.4.2    Holistic systems monitoring

Robust monitoring of software and services is painful because the signals available to detect (or even preempt) issues are so fragmented:

- Server monitoring is fed into a third-party service or a self-hosted time-series database. This data is often preaggregated or presampled because of storage and network concerns.
- Application log messages are written to the application servers as log files and hopefully collected before the server is killed or shut down.
- Customer events are sent to a third-party service and frequently not available for granular customer- or instance-level inspection.

With a unified log, any systems issue can be investigated by exploring any of the event stream data held in the unified log. The data does not have to be stored in the unified log specifically for systems-monitoring purposes; the systems administrator can explore any of that data to identify correlations, perform root cause analysis, and so on. See figure 1.13 for an example of holistic systems monitoring for a mobile-app business.

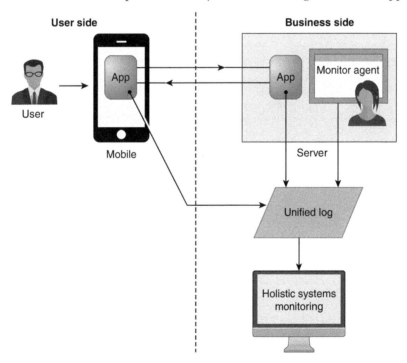

**Figure 1.13    The unified log can power holistic systems monitoring for a business operating a mobile app with a client-server architecture. Events are sent to the unified log from the mobile client, the server application, and the server's monitoring agent; if any problem occurs, the systems administrator can work with the application developers to identify the issue by looking at the whole event stream.**

### 1.4.3   *Hot-swapping data application versions*

We mentioned earlier that a unified log is readable by multiple applications simultaneously, and that each application can read events at its own pace. Each application using the unified log can independently keep track of which events it has already processed, and which are next to process.

If we can have multiple applications reading from the unified log, then it follows that we can also have multiple versions of the *same* application processing events from the unified log. This is hugely useful, as it allows us to *hot swap* our data processing applications—to upgrade our applications without taking them offline. While the current version of our application is still running, we can do the following:

1  Kick off the new version of the application from the start of the unified log
2  Let it catch up to the same cursor position as the old version
3  Switch our users over to the new version of the application
4  Shut down the old version

Figure 1.14 illustrates this hot swapping of our old application version with our new version.

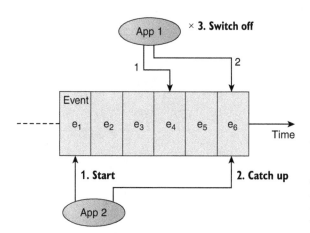

Figure 1.14   The unified log allows us to hot swap two versions of the same application. First we kick off the new application version processing from the start of our unified log, then we let the new version catch up to the old version's cursor position, and finally we switch off the old version.

The ability for each application (or application version) to maintain its own cursor position is incredibly powerful. In addition to upgrading live data processing applications without service interruptions, we can also use this capability to do the following:

- Test new versions of our applications against the live event stream
- Compare the results of different algorithms against the same event stream
- Have different users consume different versions of our applications

## *Summary*

- An event is anything that we can observe occurring at a particular point in time.
- A continuous event stream is an unterminated succession of individual events, ordered by the point in time at which each event occurred.
- Many software systems are heavily influenced by the idea of event streams, including application-level logging, web analytics, and publish/subscribe messaging.
- In the classic era of data processing, businesses operated a disparate set of on-premises systems feeding into a data warehouse. These systems featured high data latency, heavily siloed data, and many point-to-point connections between systems.
- In the hybrid era of data processing, businesses operate a hodgepodge of transactional and analytics systems. There are disparate data silos, but also attempts at "log everything" approaches with Hadoop and/or systems monitoring.
- The unified log era proposes that businesses restructure around an append-only log to which we write all events generated by our applications; software systems should communicate with each other in an uncoupled way through the unified log.
- Use cases for this unified architecture include customer feedback loops, holistic systems monitoring, and hot swapping data application versions.

# The unified log

2

**This chapter covers**

- Understanding the key attributes of a unified log
- Modeling events using JSON
- Setting up Apache Kafka, a unified log
- Sending events to Kafka and reading them from Kafka

The previous chapter introduced the idea of *events* and *continuous streams of events* and showed that many familiar software platforms and tools have event-oriented underpinnings. We recapped the history of business intelligence and data analytics, before introducing an event-centric data processing architecture, built around something called a unified log. We started to show the *why* of the unified log with some use cases but stopped short of explaining *what* a unified log is.

In this chapter, we will start to get hands-on with unified log technology. We will take a simple Java application and show how to update it to send events to a unified log. Understanding the theory and design of unified logs is important too, so we'll introduce the core attributes of the unified log first.

We have a few unified log implementations to choose from. We'll pick Apache Kafka, an open source, self-hosted unified log to get us started. With the scene set, we will code up our simple Java application, start configuring Kafka, and then code the integration between our app and Kafka. This process has a few discrete steps:

1 Defining a simple format for our events
2 Setting up and configuring our unified log
3 Writing events into our unified log
4 Reading events from our unified log

## 2.1 Understanding the anatomy of a unified log

All this talk of a unified log, but what exactly is it? A *unified log* is an append-only, ordered, distributed log that allows a company to centralize its continuous event streams. That's a lot of jargon in one sentence. Let's unpack the salient points in the following subsections.

### 2.1.1 Unified

What does it mean that our log is *unified?* It means that we have a single deployment of this technology in our company (or division or whatever), with multiple applications sending events to it and reading events from it. The Apache Kafka project (Kafka is a unified log) explains it as follows on its homepage (https://kafka.apache.org/):

> *Kafka is designed to allow a single cluster to serve as the central data backbone for a large organization.*

Having a single unified log does *not* mean that all events have to be sent to the same event stream—far from it: our unified log can contain many distinct continuous streams of events. It is up to us to define exactly how we map our business processes and applications onto continuous event streams within our unified log, as we will explore further in chapter 3.

Let's imagine a metropolitan taxi firm that is embracing the unified log whole-heartedly. Several interesting "actors" are involved in this taxi business:

- Customers booking taxis
- Taxis generating location, speed, and fuel-consumption data
- The Dispatch Engine assigning taxis to customer bookings

Figure 2.1 demonstrates one possible way of architecting this taxi firm around its new unified log implementation. The three streams share the same unified log; there is no reason for them not to. Applications such as the Dispatch Engine can read from two continuous streams of events and write into another stream.

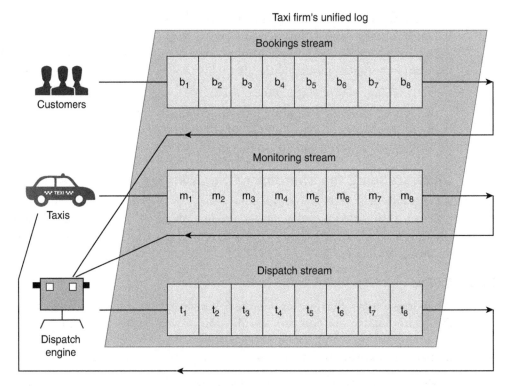

**Figure 2.1   The unified log for our taxi firm contains three streams: a taxi-booking stream, a taxi-monitoring stream, and a taxi-dispatching stream.**

### 2.1.2   *Append-only*

*Append-only* means that new events are appended to the front of the unified log, but existing events are never updated in place after they're appended. What about deletion? Events are automatically deleted from the unified log when they age beyond a configured time window, but they cannot be deleted in an ad hoc fashion. This is illustrated in figure 2.2.

Being append-only means that it is much easier for your applications to reason about their interactions with the unified log. If your application has read events up to and including event number 10, you know that you will never have to go back and look at events 1 through 10 again.

Of course, being append-only brings its own challenges: if you make a mistake when generating your events, you cannot simply go into the unified log and apply changes to fix those events, as you might in a conventional relational or NoSQL database. But we can compensate for this limitation by carefully modeling our events, building on our understanding of events from chapter 1; we'll look at this in much more detail in chapter 5.

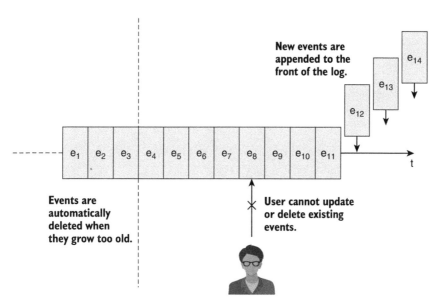

**Figure 2.2   New events are appended to the front of the log, while older events are automatically deleted when they age beyond the time window supported by the unified log. Events already in the unified log cannot be updated or deleted in an ad hoc manner by users.**

### 2.1.3  *Distributed*

*Distributed* might sound a little confusing: is the log unified, or is it distributed? Actually, it's both! *Distributed* and *unified* are referring to different properties of the log. The log is unified because a single unified log implementation is at the heart of the business, as explained previously in section 2.1.1. The unified log is *distributed* because it lives across a cluster of individual machines.

Clustered software tends to be more complex to set up, run, and reason about than software that lives inside one machine. Why do we distribute the log across a cluster? For two main reasons:

- *Scalability*—Having the unified log distributed across a cluster of machines allows us to work with event streams larger than the capacity of any single machine. This is important because any given stream of events (for example, taxi telemetry data from our previous example) could be very large. Distribution across a cluster also makes it easy for each application reading the unified log to be clustered.
- *Durability*—A unified log will replicate all events within the cluster. Without this event distribution, the unified log would be vulnerable to data loss.

To make it easier to work across a cluster of machines, unified logs tend to divide the events in a given event stream into multiple *shards* (sometimes referred to as *partitions*);

each shard will be replicated to multiple machines for durability, but one machine will be elected leader to handle all reads and writes. Figure 2.3 depicts this process.

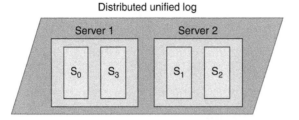

**Figure 2.3   Our unified log contains a total of four *shards* (aka *partitions*), split across two physical servers. For the purposes of this diagram, we show each partition only on its leader server. In practice, each partition would be replicated to the other server for failover.**

### 2.1.4   *Ordered*

*Ordered* means that the unified log gives each event in a shard a sequential ID number (sometimes called the *offset*) that uniquely identifies each message within the shard. Keeping the ordering restricted to the shard keeps things much simpler—because there is no need to maintain and share a global ordering across the whole cluster. Figure 2.4 shows an example of the ordering within a three-shard stream.

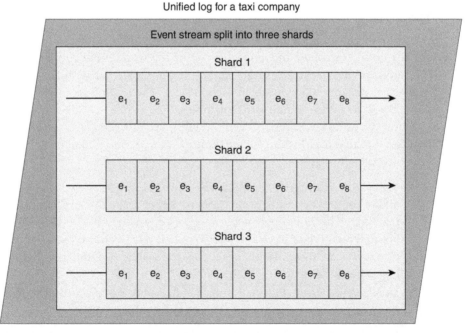

**Figure 2.4   Our taxi company's unified log contains a single stream, holding events generated by customers, taxis, and the dispatching engine. In this case, the events are split into three distinct shards. Each shard maintains its own ordering from e1 through to e8.**

This ordering gives the unified log much of its power: different applications can each maintain their own cursor position for each shard, telling them which events they have already processed, and thus which events they should process next.

If a unified log did not order events, each consuming application would have to do one of the following:

- *Maintain an event manifest*—Keep a list of all event IDs processed so far and share this with the unified log to determine which events to retrieve next. This approach is conceptually similar to maintaining a manifest of processed files in traditional batch processing.
- *Update or even delete events in the log*—Set some kind of flag against events that have been processed to determine which events to retrieve next. This approach is similar to "popping" a message off a first in, first out (FIFO) queue.

Both of these alternatives would be extremely painful. In the first case, the number of event IDs to keep track of would become hugely unwieldy, reminiscent of the Jorge Luis Borges short story:

> In that Empire, the Art of Cartography attained such Perfection that the map of a single Province occupied the entirety of a City, and the map of the Empire, the entirety of a Province.

> —Jorge Luis Borges, "Del Rigor en la Ciencia" (1946)

The second option would not be much better. We would lose the immutability of our unified log and make it hard for multiple applications to share the stream, or for the sample application to "replay" events it had already processed. And in both situations, the unified log would have to support random access to individual events from consuming applications. So, ordering the log makes a huge amount of sense.

To recap: you have now seen why the unified log is *unified*, why it is *append-only*, why it is *ordered*, and why, indeed, it is *distributed*. Hopefully, this has started to clarify how a unified log is architected, so let's get started using one.

## 2.2 *Introducing our application*

With the basic theory of the unified log set out, let's begin to put the theory to work! In each part of this book, we will be working with a fictitious company that wants to implement a unified log across its business. To keep things interesting, we will choose a company in a different sector each time. Let's start with a sector that almost all of us have experienced, at least as customers: e-commerce.

Imagine that we work for a sells-everything e-commerce website; let's call it *Nile*. The management team at Nile wants the company to become much more dynamic and responsive: Nile's analysts should have access to up-to-the-minute sales data, and Nile's systems should react to customer behavior in a timely fashion. As you will see in this part of the book, we can meet their requirements by implementing a unified log.

### 2.2.1   *Identifying our key events*

Online shoppers browse products on the Nile website, sometimes adding products to their shopping cart, and sometimes then going on to buy those products through the online checkout. Visitors can do plenty of other things on the website, but Nile's executives and analysts care most about this Viewing through Buying workflow. Figure 2.5 shows a typical shopper (albeit with somewhat eclectic shopping habits) going through this workflow.

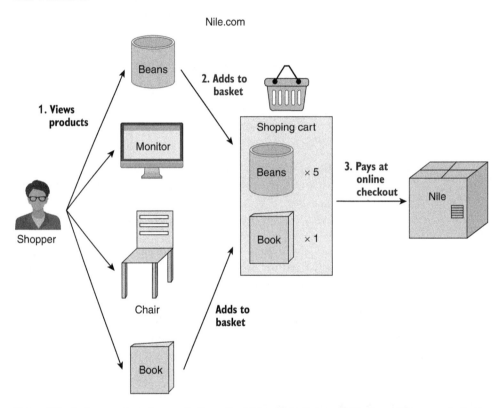

**Figure 2.5   A shopper views four products on the Nile website before adding two of those products to the shopping cart. Finally, the shopper checks out and pays for those items.**

Even though our standard definition of an event as *subject-verb-object* will be described in section 2.1.7, we can already identify three discrete events in this Viewing through Buying workflow:

1   *Shopper* views *product* at *time*—Occurs every time the shopper views a product, whether on a product's dedicated Detail page, or on a general Catalog page that happens to include the product.

2   *Shopper* adds *item* to *cart* at *time*—Occurs whenever the shopper adds one of those products to the shopping basket. A product is added to the basket with a quantity of one or more attached.

**3** *Shopper* places *order* at *time*—Occurs when the shopper checks out, paying for the items in the shopping basket.

To avoid complicating our lives later, this part of the book keeps this workflow deliberately simple, steering clear of more-complex interactions, such as the shopper adjusting the quantity of an item in the shopping basket, or removing an item from the basket at checkout. But no matter: the preceding three events represent the essence of the shopping experience at Nile.

### 2.2.2 Unified log, e-commerce style

Nile wants to introduce Apache Kafka (https://kafka.apache.org) to implement a unified log across the business. Future chapters cover Kafka in much more detail. For now, it's important to understand only that Kafka is an open source unified log technology that runs on the Java virtual machine (JVM).

We can define an initial event stream (aka *Kafka topic*) to record the events generated by shoppers. Let's call this stream `raw-events`. In figure 2.6, you can see our

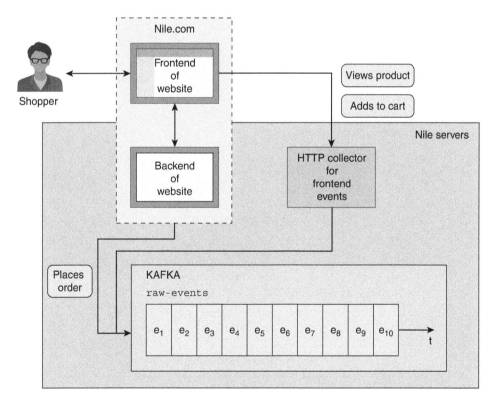

**Figure 2.6** Our three types of event are flowing through, into the `raw-events` topic in Kafka. Order events are written to Kafka directly from the website's backend; the in-browser events of viewing products and adding products to the cart are written to Kafka via an HTTP server that collects events.

three types of events feeding into the stream. To make this a little more realistic, we've made a distinction based on how the events are captured:

- *Browser-generated events*—The *Shopper views product* and *Shopper adds item to basket* events occur in the user's browser. Typically, there would be some JavaScript code to send the event to an HTTP-based event collector, which would in turn be tasked with writing the event to the unified log.[1]
- *Server-side events*—A valid *Shopper places order* event is confirmed server-side only after the payment processing has completed. It is the responsibility of the web server to write this event to Kafka.

What should our three event types look like? Unified logs don't typically have an opinion on the schema for the events passing through them. Instead they treat each event as a "sealed envelope," a simple array of bytes to be appended to the log and shared as is with applications when requested. It is up to us to define the internal format of our events—a process we call *modeling* our events.

The rest of this chapter focuses on just a single Nile event type: *Shopper views product at time*. We will come back to the other two Nile event types in the next two chapters.

### 2.2.3   *Modeling our first event*

How should we model our first event, *Shopper views product at time*? The secret is to realize that our event already has an inherent structure: it follows the grammatical rules of a sentence in the English language. For those of us who are a little rusty on English language grammar, figure 2.7 maps out the key grammatical components of this event, expressed as a sentence.

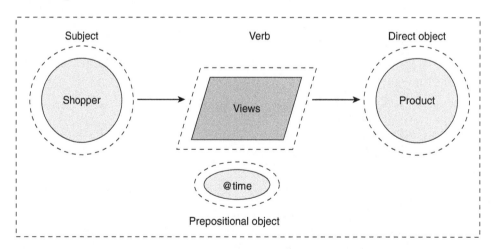

**Figure 2.7   Our *shopper* (subject) *views* (verb) a *product* (direct object) at *time* (prepositional object).**

---

[1]   This JavaScript code can be found at https://github.com/snowplow/snowplow-javascript-tracker.

Let's go through each of these components in turn:

- The "shopper" is the sentence's *subject*. The subject in a sentence is the entity carrying out the action: *Jane* views an iPad at midday.
- "views" is the sentence's *verb*, describing the action being done by the *subject*: "Jane *views* an iPad at midday."
- The "product" being viewed is the *direct object*, or simply *object*. This is the entity to which the action is being done: "Jane views *an iPad* at midday."
- The time of the event is, strictly speaking, another object—an *indirect object*, or preposition*al object*, where the preposition is "at": "Jane views an iPad at *midday.*"

We now have a way of breaking our event into its component parts, but so far this description is only human-readable. A computer couldn't easily parse it. We need a way of formalizing this structure further, ideally into a *data serialization format* that is understandable by humans but also can be parsed by computers.

We have lots of options for serializing data. For this chapter, we will pick JavaScript Object Notation (JSON). JSON has the attractive property of being easily written and read by both people and machines. Many, if not most, developers setting out to model their company's continuous event streams will start with JSON.

The following listing shows a possible representation for our *Shopper views product at time* event in JSON.

**Listing 2.1 shopper_viewed_product.json**

```
{
  "event": "SHOPPER_VIEWED_PRODUCT",         ◁— The event type as a simple
  "shopper": {                        ◁—         string with a verb in the
    "id": "123",                                 simple past tense
    "name": "Jane",                        An object representing the
    "ipAddress": "70.46.123.145"           shopper who viewed the product
  },
  "product": {           ◁—  An object representing the
    "sku": "aapl-001",         product that was viewed
    "name": "iPad"
  },
  "timestamp": "2018-10-15T12:01:35Z"   ◁—  When the event occurred as an
}                                            ISO 8601-compatible timestamp
```

Our representation of this event in JSON has four properties:

- event holds a string representing the type of event.
- shopper represents the person (in this case, a woman named Jane) viewing the product. We have a unique id for the shopper, her name and a property called ipAddress, which is the IP address of the computer she is browsing on.
- product contains the sku (stock keeping unit) and name of the product, an iPad, being viewed.
- timestamp represents the exact time when the shopper viewed the product.

To look at it another way, our event consists of two pieces of *event metadata* (namely, the event and the `timestamp`), and two *business entities* (the `shopper` and the `product`). Now that you understand the specific format of our event in JSON, we need somewhere to send them!

## 2.3 *Setting up our unified log*

We are going to send the event stream generated by Nile into a unified log. For this, we're going to pick Apache Kafka. Future chapters cover Kafka in much more detail. For now, it's just important to understand that Kafka is an open source (Apache License 2.0) unified log technology that runs on the JVM.

Be aware that we are going to start up *and leave running* multiple pieces of software in the next subsection. Get ready with a few terminal windows (or a tabbed terminal client if you're lucky).

### 2.3.1 *Downloading and installing Apache Kafka*

A Kafka cluster is a powerful piece of technology. But, fortunately, it's simple to set up a cheap-and-cheerful single-node cluster just for testing purposes. First, download Apache Kafka version 2.0.0:

```
http://archive.apache.org/dist/kafka/2.0.0/kafka_2.12-2.0.0.tgz
```

You will have to access that link in a browser. You cannot use `wget` or `curl` to download the file directly. When you have it, un-tar it:

```
$ tar -xzf kafka_2.12-2.0.0.tgz
$ cd kafka_2.12-2.0.0
```

Kafka uses Apache ZooKeeper (https://zookeeper.apache.org) for cluster coordination, among other things. Deploying a production-ready ZooKeeper cluster requires care and attention, but fortunately Kafka comes with a helper script to set up a single-node ZooKeeper instance. Run the script like so:

```
$ bin/zookeeper-server-start.sh config/zookeeper.properties
[2018-10-15 23:49:05,185]
 INFO Reading configuration from: config/zookeeper.properties
 (org.apache.zookeeper.server.quorum.QuorumPeerConfig)
[2018-10-15 23:49:05,190] INFO
 autopurge.snapRetainCount set to 3
 (org.apache.zookeeper.server.DatadirCleanupManager)
[2018-10-15 23:49:05,191] INFO
 autopurge.purgeInterval set to 0
 (org.apache.zookeeper.server.DatadirCleanupManager)
...
[2018-10-15 23:49:05,269] INFO
 minSessionTimeout set to -1 (org.apache.zookeeper.server.ZooKeeperServer)
[2018-10-15 23:49:05,270] INFO
 maxSessionTimeout set to -1 (org.apache.zookeeper.server.ZooKeeperServer)
[2018-10-15 23:49:05,307] INFO
```

```
binding to port 0.0.0.0/0.0.0.0:2181
(org.apache.zookeeper.server.NIOServerCnxnFactory)
```

Now we are ready to start Kafka in a second terminal:

```
$ bin/kafka-server-start.sh config/server.properties
[2018-10-15 23:52:05,332] INFO Registered
 kafka:type=kafka.Log4jController MBean
 (kafka.utils.Log4jControllerRegistration$)
[2018-10-15 23:52:05,374] INFO starting (kafka.server.KafkaServer)
[2018-10-15 23:52:05,375] INFO
Connecting to zookeeper on localhost:2181 (kafka.server.KafkaServer)
...
[2018-10-15 23:52:06,293] INFO
Kafka version : 2.0.0 (org.apache.kafka.common.utils.AppInfoParser)
[2018-10-15 23:52:06,337] INFO
Kafka commitId : 3402a8361b734732
      (org.apache.kafka.common.utils.AppInfoParser)
[2018-10-15 23:52:06,411] INFO
[KafkaServer id=0] started (kafka.server.KafkaServer)
```

Great, we now have both ZooKeeper and Kafka running. Our bosses at Nile will be pleased.

### 2.3.2 *Creating our stream*

Kafka doesn't use our exact language of continuous event streams. Instead, Kafka producers and consumers interact with *topics*; you might remember the language of topics from our publish/subscribe example with the NSQ message queue in chapter 1.

Let's create a new topic in Kafka called `raw-events`:

```
$ bin/kafka-topics.sh --create --topic raw-events \
  --zookeeper localhost:2181 --replication-factor 1 --partitions 1
Created topic "raw-events".
```

Let's briefly go through the second line of arguments:

- The `--zookeeper` argument tells Kafka where to find the ZooKeeper that is keeping track of our Kafka setup.
- The `--replication-factor` of 1 means that the events in our topic will not be replicated to another server. In a production system, we would increase the replication factor so that we can continue processing in the face of server failures.
- The `--partitions` setting determines how many shards we want in our event stream. One partition is plenty for our testing.

We can see our new topic if we run the `list` command:

```
$ bin/kafka-topics.sh --list --zookeeper localhost:2181
raw-events
```

If you don't see `raw-events` listed or get some kind of Connection Refused error, go back to section 2.3.1 and run through the setup steps in the exact same order again.

### 2.3.3   *Sending and receiving events*

Now we are ready to send our first events into the raw-events topic in Nile's unified
log in Kafka. We can do this at the command line with a simple producer script. Let's
start it running:

```
$ bin/kafka-console-producer.sh --topic raw-events \
  --broker-list localhost:9092

[2018-10-15 00:28:06,166] WARN Property topic is not valid
 (kafka.utils.VerifiableProperties)
```

Per the command-line arguments, we will be sending, or *producing*, events to the Kafka
topic raw-events, which is available from the Kafka broker available on our local server
on port 9092. To send in our events, you type them in and press Enter after each one:

```
{ "event": "SHOPPER_VIEWED_PRODUCT", "shopper": { "id": "123", "name":
"Jane", "ipAddress": "70.46.123.145" }, "product": { "sku": "aapl-001",
"name": "iPad" }, "timestamp": "2018-10-15T12:01:35Z" }
{ "event": "SHOPPER_VIEWED_PRODUCT", "shopper": { "id": "456", "name":
Mo", "ipAddress": "89.92.213.32" }, "product": { "sku": "sony-072", "name":
"Widescreen TV" }, "timestamp": "2018-10-15T12:03:45Z" }
{ "event": "SHOPPER_VIEWED_PRODUCT", "shopper": { "id": "789", "name":
"Justin", "ipAddress": "97.107.137.164" }, "product": { "sku": "ms-003",
"name": "XBox One" }, "timestamp": "2018-10-15T12:05:05Z" }
```

Press Ctrl-D to exit. We have sent in three *Shopper views product* events to our Kafka
topic. Now let's read the same events out of our unified log, using the Kafka command-
line consumer script:

```
$ bin/kafka-console-consumer.sh --topic raw-events --from-beginning \
  --bootstrap-server localhost:9092
{ "event": "SHOPPER_VIEWED_PRODUCT", "shopper": { "id": "123",
 "name": "Jane", "ipAddress": "70.46.123.145" }, "product": { "sku":
 "aapl-001", "name": "iPad" }, "timestamp": "2018-10-15T12:01:35Z" }
{ "event": "SHOPPER_VIEWED_PRODUCT", "shopper": { "id": "456",
 "name": "Mo", "ipAddress": "89.92.213.32" }, "product": { "sku":
 "sony-072", "name": "Widescreen TV" }, "timestamp":
 "2018-10-15T12:03:45Z" }
{ "event": "SHOPPER_VIEWED_PRODUCT", "shopper": { "id": "789",
 "name": "Justin", "ipAddress": "97.107.137.164" }, "product": {
 "sku": "ms-003", "name": "XBox One" }, "timestamp":
 "2018-10-15T12:05:05Z" }
```

Success! Let's go through these arguments briefly:

- We are specifying raw-events as the topic we want to read or consume events
  from.
- The argument --from-beginning indicates that we want to consume events
  from the start of the stream onward.
- The --bootstrap-server argument tells Kafka where to find the running
  Kafka broker.

This time, press Ctrl-C to exit. As a final test, let's pretend that Nile has a second application that also wants to consume from the `raw-events` stream. It's a key property of a unified log technology such as Kafka that we can have multiple applications reading from the same event stream at their own pace. Let's simulate this with another call of the consumer script:

```
$ bin/kafka-console-consumer.sh --topic raw-events --from-beginning \
  --bootstrap-server localhost:9092
{ "event": "SHOPPER_VIEWED_PRODUCT", "shopper": { "id": "123",
 "name": "Jane", "ipAddress": "70.46.123.145" }, "product": { "sku":
 "aapl-001", "name": "iPad" }, "timestamp": "2018-10-15T12:01:35Z" }
{ "event": "SHOPPER_VIEWED_PRODUCT", "shopper": { "id": "456",
 "name": "Mo", "ipAddress": "89.92.213.32" }, "product": { "sku":
 "sony-072", "name": "Widescreen TV" }, "timestamp":
 "2018-10-15T12:03:45Z" }
{ "event": "SHOPPER_VIEWED_PRODUCT", "shopper": { "id": "789",
 "name": "Justin", "ipAddress": "97.107.137.164" }, "product": { "sku":
 "ms-003", "name": "XBox One" }, "timestamp": "2018-10-15T12:05:05Z" }
^CConsumed 3 messages
```

Fantastic—our second request to read the `raw-events` topic from the beginning has returned the exact same three events. This helps illustrate the fact that Kafka is serving as a kind of *event stream database*. Compare this to a pub/sub message queue, where a single subscriber reading messages "pops them off the queue," and they are gone for good.

We can now send a well-structured event stream to Apache Kafka—a simple event stream for Nile's data engineers, analysts, and scientists to work with. In the next chapter, we will start to make this raw event stream even more useful for our coworkers at Nile by performing simple transformations on the events in-stream.

## Summary

- A unified log is an append-only, ordered, distributed log that allows a company to centralize its continuous event streams.
- We can generate a continuous stream of events from our applications and send those events into an open source unified log technology such as Apache Kafka.
- We represented our events in JSON, a widely used data serialization format, by using a simple structure that echoes the English-language grammar structure of our events.
- We created a topic in Kafka to store our Nile events—a *topic* is Kafka-speak for a specific stream of events.
- We wrote, or *produced*, events to our Kafka topic by using a simple command-line script.
- We read, or *consumed*, events from our Kafka topic by using a simple command-line script.
- We can have multiple consumers reading from the same Kafka topic at their own pace, which is a key building block of the new unified log architecture.

# Event stream processing with Apache Kafka

**This chapter covers**

- Introducing event stream processing
- Writing applications that process individual events
- Validating and enriching events
- Writing enriched events to output streams

In the preceding chapter, we focused on getting a stream of well-structured events from Nile, our fictitious e-commerce retailer, into our unified log, Apache Kafka. Now that we have a continuous stream of events flowing into our unified log, what can we do with these events? We can *process* them.

At its simplest, *event processing* involves reading one or more events from an event stream and doing something to those events. That processing operation could be filtering an event from the stream, validating the event against a schema, or enriching the event with additional information. Or we could be processing multiple events at a time, perhaps with a view to reordering them or creating some kind of summary or aggregate of those events.

This chapter introduces event stream processing briefly before jumping into a concrete example, processing the Nile event stream. Our new stream-processing application will treat Nile's raw event stream as its own input stream, and it will then generate an output event stream based on those incoming events. You'll see how, by reading one stream from Kafka and writing another stream back into Kafka, we are able to use our unified log as a kind of "superglue" between our business's different apps.

We'll keep our stream-processing application simple: we'll stick to *validating* Nile's incoming raw events and *enriching* the valid events. *Enriching* means adding interesting extra information to an event. For a relatively simple example in this chapter, we will enrich the events with the customer's geographical location, using the MaxMind geo-location database.

Let's get started.

## 3.1 Event stream processing 101

In chapter 2, we defined an initial event type for Nile, set up Apache Kafka locally, and then went about sending those events into Kafka. All of this initial plumbing was a means to an end; the end is processing the Nile event stream now available in Kafka. But *event stream processing* is not a widely known term, so let's take a brief look at why we process event streams, and what that entails.

### 3.1.1 Why process event streams?

Your business might want to process event streams for a variety of reasons. Perhaps you want to do one of the following:

- Back up the events to long-term storage such as HDFS or Amazon S3
- Monitor the event stream for patterns or abnormalities and send alerts when these are detected
- "Drip-feed" the events into databases such as Amazon Redshift, Vertica, Elasticsearch, or Apache Cassandra
- Derive new event streams from the original event stream—for example, filtered, aggregated, or enriched versions of the original event stream

Figure 3.1 illustrates all four of these use cases.

The general term for what all of these example applications are doing is *event stream processing*. We can say that any program or algorithm that understands the time-ordered, append-only nature of a continuous event stream and can consume events from this stream in a meaningful way is able to *process* this stream.

Fundamentally, only two types of processing can be performed on a single continuous event stream:

- *Single-event processing*—A single event in the event stream will produce zero or more output data points or events.
- *Multiple-event processing*—Multiple events from the event stream will collectively produce zero or more output data points or events.

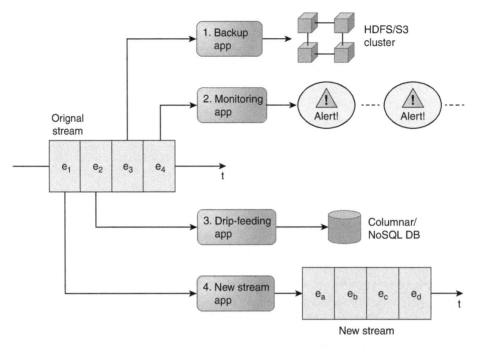

**Figure 3.1**  Our original event stream is being processed by four applications, one of which is generating a new event stream of its own. Note that each of our four applications can have a different current offset, or *cursor position*, in the original stream; this is a feature of unified logs.

## Complex event processing

You may hear people talk about *complex event processing* (CEP) and wonder how this relates to event stream processing as described in this chapter. In fact, Manning has a 2011 book on CEP called *Event Processing in Action* by Opher Etzion and Peter Niblett.

As far as I can tell, CEP emphasizes the derivation of "complex events" from simpler input events, although this is an important use case for our event stream processing approaches as well. A more significant difference is that CEP thinking predates unified log technologies like Apache Kafka, so CEP systems will tend to work on much smaller (and potentially unordered) event streams than we will look at.

Another difference is that the CEP ecosystem seems to be dominated by commercial applications with drag-and-drop graphical user interfaces and/or declarative event query languages. By contrast, event stream processing as we define it is much more programmer-focused, with most algorithms being hand-rolled in Java, Scala, Python, or similar.

Of the various CEP products introduced in *Event Processing in Action*, the only one I have encountered being used in a modern event stream processing context is Esper (www.espertech.com/esper), an open-source CEP tool with its own event query language.

Figure 3.2 illustrates both types. We distinguish between these two types of stream processing because they differ hugely from each other in terms of complexity. Let's look at each in turn.

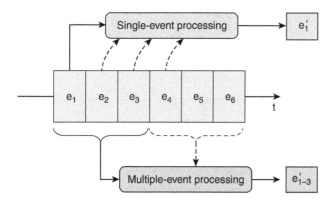

**Figure 3.2  Our single-event processing app works on only one event from the source stream at a time. By contrast, the application at the bottom is reading three events from the source stream in order to generate a single output event.**

### 3.1.2  *Single-event processing*

The first case, *single-event processing*, is straightforward to implement: we read the next event from our continuous event stream and apply some sort of transformation to it. We can apply many transformations, and common ones include the following:

- *Validating* the event—Checking, for example, "Does this event contain all the required fields?"
- *Enriching* the event—Looking up, for example, "Where is this IP address located?"
- *Filtering* the event—Asking, for example, "Is this error critical?"

We could also apply a combination of these. Many of these possible transformations would generate either zero or one data points or events, but equally they could produce multiple data points or events. For example, a process could produce a stream of validation warnings as well as a stream of enriched events and filter out some events entirely. This is illustrated in figure 3.3.

Regardless of the transformations we attempt, in the single event case, any stream processing is conceptually simple, because we have to act on only a single event at a time. Our stream processing application can have the memory of a goldfish: no event matters except the one being read right now.

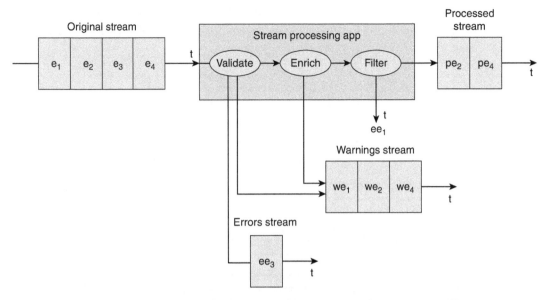

**Figure 3.3 Here the stream processing application is validating, enriching, and filtering an incoming raw stream. Events that make it through the whole transformation are added to our processed stream. Events that fail validation are added to our errors stream. Transformation warnings are added to our warnings stream.**

### 3.1.3 Multiple-event processing

In *multiple-event processing*, we have to read multiple events from the event stream in order to generate some kind of output. Plenty of algorithms and queries fit into this pattern, including these:

- *Aggregating*—Applying aggregate functions such as minimum, maximum, sum, count, or average on multiple events
- *Pattern matching*—Looking for patterns or otherwise summarizing the sequence, co-occurrence, or frequency of multiple events
- *Sorting*—Reordering events based on a sort key

Figure 3.4 features a trio of stream processing apps; each is working on multiple events and applying multiple event algorithms and queries.

Processing multiple events at a time is significantly more complex conceptually and technically than processing single events. Chapter 4 explores the processing of multiple events at a time in much more detail.

## 3.2 Designing our first stream-processing app

Let's return to Nile, our fictitious e-commerce retailer. Chapter 2 introduced Nile's three event types, and we represented one of these event types, *Shopper views product*, in JSON. We wrote these JSON events into a Kafka topic called `raw-events`, and then read these events out again from the same topic. In this chapter, we will go further and start to do some *single-event processing* on this event stream. Let's get started!

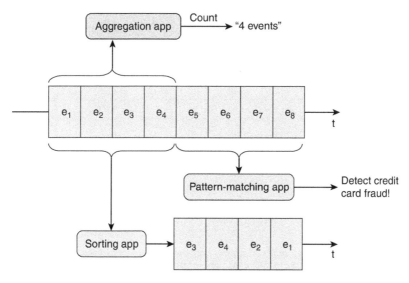

**Figure 3.4  We have three apps processing multiple events at a time: an Aggregation app, which is counting events; a Pattern Matching app, which is looking for event patterns indicative of credit card fraud; and, finally, a Sorting app, which is reordering our event stream based on a property of each event.**

### 3.2.1  Using Kafka as our company's glue

The data scientists at Nile want to start by analyzing one of our three event types: the *Shopper views product* events. There is just one problem: the Data Science team at Nile is split by geography, and each subteam wants to analyze only shopper behaviors from specific countries.

We have been asked to build a stream processing application to do the following:

1  Read the `raw-events` topic in Kafka
2  Figure out where each shopper is located
3  Write the events, now with the country and city attached, out to another Kafka topic

Figure 3.5 illustrates this flow.

This example shows how we can start to use Apache Kafka, our company's unified log, as the "glue" between systems without getting, well, stuck. By simply processing the incoming stream as requested and writing the new events back to further Kafka topics, we don't have to know anything about how either team of data scientists will work with the events.

After we have agreed on a format for the events with the data scientists, we can then leave them to work with the new event stream however they want. They can write their own stream processing applications, or store all the events in an analytics database, or archive the events in Hadoop and write machine learning or graph algorithms to use on them; it doesn't matter to us. To overload this part of the book with

Nile's unified log (Kafka)

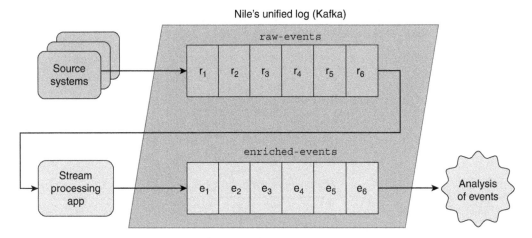

**Figure 3.5  Our first stream-processing app will read events from the `raw-events` topic in Apache Kafka and write enriched events back to a new topic in Kafka. As our unified log, Kafka is the glue between multiple applications.**

metaphors, our unified log is acting as the Esperanto for our different applications and users.

### 3.2.2  Locking down our requirements

Before writing any code, we need to bottom out the requirements for our stream processing app. Remember that the *Shopper views product* events occur in the shopper's web browser and are relayed to Kafka via some kind of HTTP-based event collector. The events are created in an environment outside our direct control, so the first step is to *validate* that each event found in `raw-events` has the expected structure. We want to protect the Nile data scientists from any defective events; they are paid too much to spend their time cleaning up bad data!

After we have validated our events, we need to identify where each event originated geographically. How can we determine where our Nile shoppers are located? Let's look back at the data points in each incoming *Shopper views product* event:

```
{"event": "SHOPPER_VIEWED_PRODUCT", "shopper": {"id": "123",
 "name": "Jane", "ipAddress": "70.46.123.145"}, "product": {"sku":
 "aapl-001", "name": "iPad"}, "timestamp": "2018-10-15T12:01:35Z" }
```

We are in luck: each of our events includes the IP address of the computer that our shopper is using. A company called MaxMind (www.maxmind.com) provides a free-to-use database that maps IP addresses to geographical location. We can look up each shopper's IP address in the MaxMind geo-IP database to determine where the shopper is located at that point in time. When we use algorithms or external databases to add extra data points to an event, we typically say that we are *enriching* the event.

So far, we are validating the incoming event and then enriching it. The final step will be to write out the validated, enriched events to a new Kafka topic: `enriched-events`. Our work is then done: the Nile data science teams will read the events from those topics and perform whatever analysis they want.

Putting it together, we need to create a stream processing application that does the following:

- *Reads* individual events from our Kafka topic `raw-events`
- *Validates* the event's IP address, sending any validation failures to a dedicated Kafka topic, called `bad-events`
- *Enriches* our validated events with the geographical location of the shopper by using the MaxMind geo-IP database
- *Writes* our validated, enriched events to the `enriched-events` Kafka topic

We can now put together a more detailed diagram for the stream processing application we are going to build. Figure 3.6 provides the specifics.

We are now ready to start building our stream processing app!

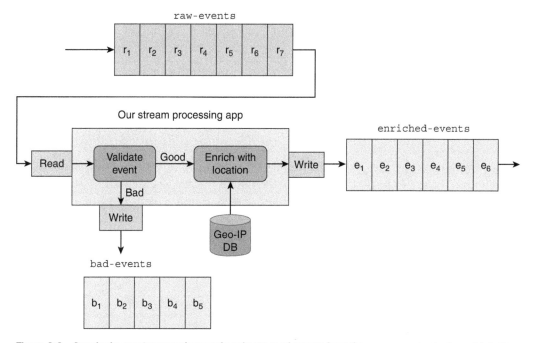

**Figure 3.6**  Our single-event processing app is going to read events from the `raw-events` topic, validate the incoming events, enrich the valid events with the geographical location, and route the enriched events to `enriched-events`. Any errors will be written out to `bad-events`.

## 3.3    Writing a simple Kafka worker

To keep things simple, we will make two passes through our stream processing app:

1  We will create a simple *Kafka worker*, which can read from our raw-events topic in Kafka and write all events to a new topic.

2  We will evolve our Kafka worker into a complete *single-event processor*, which handles validation, enrichment, and routing as per Nile's requirements.

Let's get started on the first pass.

### 3.3.1    Setting up our development environment

Our first cut of the stream processing app will let us get comfortable reading and writing to Kafka topics from Java, without Nile's pesky business logic getting in the way. We chose Java (version 8) because it has first-class support for Kafka and should be a familiar programming language for most readers; however, you won't need to be a Java guru to follow along. For our build tool, we will use Gradle, which is growing in popularity as a friendlier and less verbose (but still powerful) alternative to Ant and Maven.

Let's set up our development environment. First, you should download and install the latest Java SE8 JDK from here:

```
www.oracle.com/technetwork/java/javase/downloads/index.html
```

Next you need to download and install Gradle from here:

```
www.gradle.org/downloads
```

In order to avoid the process of manually installing every library and framework required for the examples in this book, we will take advantage of Vagrant, which provides an easy-to-configure, reproducible, and portable work environment. Using Vagrant, you can quickly install and manage a virtual machine environment with everything you need to run the examples. We selected Vagrant because it requires little effort to set up and use. In production, you might choose another tool, such as Docker, that could serve a similar purpose. If you're unfamiliar with Vagrant, you can visit www.vagrantup.com to get started.

> **For Vagrant users**
>
> If you are a Vagrant user (or would like to become one), you are in luck: we have created a Vagrant-based development environment for this book, using Ansible to install all required dependencies into a 64-bit Ubuntu 16.04.5 LTS (Xenial Xerus).
>
> If you haven't already, install Vagrant (www.vagrantup.com) and VirtualBox (www.virtualbox.org).

You can then check the environment out of GitHub like so:

```
$ git clone https://github.com/alexanderdean/Unified-Log-Processing.git
```

Now start the environment:

```
$ cd Unified-Log-Processing
$ vagrant up && vagrant ssh
```

And that's it! You can now browse to a specific chapter's example code and build it:

```
$ cd ch03/3.3
$ gradle jar
```

Be aware that the Ansible step within `vagrant up` will take a long time.

Finally, let's check that everything is installed where it should be:

```
$ java -version
java version "1.8.0_181"
...
$ gradle -v
...
Gradle 4.10.2
...
```

All present and correct? Now we are ready to create our application.

### 3.3.2  Configuring our application

Let's create our project, which we will call `StreamApp`, using Gradle. First, create a directory called `nile`. Then switch to that directory and run the following:

```
$ gradle init --type java-library
...
BUILD SUCCESSFUL
...
```

Gradle will create a skeleton project in that directory, containing a couple of Java source files for stub classes called Library.java and LibraryTest.java, as per figure 3.7. You can delete these two files; we'll be writing our own code shortly.

Next let's prepare our Gradle project build file. Edit the file build.gradle and replace its current contents with the following listing.

**Figure 3.7  Delete the files Library.java and LibraryTest.java from your generated Gradle project.**

---

**Listing 3.1   build.gradle**

```
plugins {
  // Apply the java-library plugin to add support for Java Library
  id 'java'
  id 'java-library'
  id 'application'
}

sourceCompatibility = '1.8'              For compatibility
                                         with Java 8 and up

mainClassName = 'nile.StreamApp'

version = '0.2.0'
                                         Our dependencies on
                                         third-party libraries
dependencies {
  compile 'org.apache.kafka:kafka-clients:2.0.0'
  compile 'com.maxmind.geoip2:geoip2:2.12.0'
  compile 'com.fasterxml.jackson.core:jackson-databind:2.9.7'
  compile 'org.slf4j:slf4j-api:1.7.25'
}

repositories {
  jcenter()                  Let's assemble
}                            StreamApp into
                             a fat jar.
jar {
  manifest {
    attributes 'Main-Class': mainClassName
  }

  from {
    configurations.compile.collect {
      it.isDirectory() ? it : zipTree(it)
    }
  } {
    exclude "META-INF/*.SF"
    exclude "META-INF/*.DSA"
    exclude "META-INF/*.RSA"
  }
}
```

Note the library dependencies we have added to our app:

- kafka-clients, for reading from and writing to Kafka
- jackson-databind, which is a library for parsing and manipulating JSON
- geoip-api, which we will use for our MaxMind geo-IP enrichment

Let's just check that we can build our new StreamApp project without issue (this may take two or three minutes):

```
$ gradle compileJava
...
BUILD SUCCESSFUL
...
```

Great—we are ready for the next step: building our Kafka event consumer.

### 3.3.3    Reading from Kafka

As a first step, we need to read individual raw events from our Kafka topic `raw-events`. In Kafka parlance, we need to write a *consumer*. Remember that in the preceding chapter, we depended on the Kafka command-line tools to write events to a topic, and to read events back out of that topic. In this chapter, we will write our own consumer in Java, using the Kafka Java client library.

Writing a simple Kafka consumer is not particularly difficult. Let's create a file for it, called src/main/java/nile/Consumer.java. Add in the code in the following listing.

**Listing 3.2    Consumer.java**

```java
package nile;

import java.util.*;

import org.apache.kafka.clients.consumer.*;

public class Consumer {

  private final KafkaConsumer<String, String> consumer;
  private final String topic;

  public Consumer(String servers, String groupId, String topic) {
    this.consumer = new KafkaConsumer<String, String>(
      createConfig(servers, groupId));
    this.topic = topic;
  }

  public void run(IProducer producer) {
    this.consumer.subscribe(Arrays.asList(this.topic));
    while (true) {
      ConsumerRecords<String, String> records = consumer.poll(100);
      for (ConsumerRecord<String, String> record : records) {
        producer.process(record.value());
      }
    }
  }

  private static Properties createConfig(String servers, String groupId) {
    Properties props = new Properties();
    props.put("bootstrap.servers", servers);
    props.put("group.id", groupId);
    props.put("enable.auto.commit", "true");
    props.put("auto.commit.interval.ms", "1000");
    props.put("auto.offset.reset", "earliest");
    props.put("session.timeout.ms", "30000");
    props.put("key.deserializer",
      "org.apache.kafka.common.serialization.StringDeserializer");
    props.put("value.deserializer",
      "org.apache.kafka.common.serialization.StringDeserializer");
```

Our Kafka consumer will read Kafka records for which the key and value are both strings.

Subscribe our consumer to the given Kafka topic.

Looping forever, fetch records from the Kafka topic.

Feed each record's value to the process method of our producer.

Identify this consumer as belonging to a specific consumer group.

```
    return props;
  }
}
```

So far, so good; we have defined a consumer that will read all the records from a given Kafka topic and hand them over to the `process` method of the supplied producer. We don't need to worry about most of the consumer's configuration properties, but note the `group.id`, which lets us associate this app with a specific Kafka *consumer group*. We could run multiple instances of our app all with the same `group.id` to share out the topic's events across all of our instances; by contrast, if each instance had a different `group.id`, each instance would get all of Nile's `raw-events`.

### 3.3.4  Writing to Kafka

See how our consumer is going to run the `IProducer.process()` method for each incoming event? To keep things flexible, the two producers we write in this chapter will both conform to the `IProducer` interface, letting us easily swap out one for the other. Let's now define this interface in another file, called src/main/java/nile/IProducer.java. Add in the code in the following listing.

> **Listing 3.3  IProducer.java**

```
package nile;

import java.util.Properties;

import org.apache.kafka.clients.producer.*;          ┐ Our abstract process
                                                      │ method, for concrete
public interface IProducer {                          │ implementations of
                                                      │ IProducer to
  public void process(String message);      ◀────────┘ instantiate

  public static void write(KafkaProducer<String, String> producer,
    String topic, String message) {                   ◀─┐ A static helper to
    ProducerRecord<String, String> pr = new ProducerRecord(  │ write a record to
      topic, message);                                       │ a Kafka topic
    producer.send(pr);
  }

  public static Properties createConfig(String servers) {   ◀─┐ A static helper
    Properties props = new Properties();                       │ to configure a
    props.put("bootstrap.servers", servers);                   │ Kafka producer
    props.put("acks", "all");
    props.put("retries", 0);
    props.put("batch.size", 1000);
    props.put("linger.ms", 1);
    props.put("key.serializer",
      "org.apache.kafka.common.serialization.StringSerializer");
    props.put("value.serializer",
      "org.apache.kafka.common.serialization.StringSerializer");
    return props;
  }
}
```

This is a great start, but for this to be useful, we need a concrete implementation of IProducer. Remember that this section of the chapter is just a warm-up: we want to pass the incoming raw-events into a second topic with the events themselves untouched. We now know enough to implement a simple *pass-through producer*, by adding the code in the following listing into a new file called src/main/java/nile/PassthruProducer.java.

**Listing 3.4  PassthruProducer.java**

```
package nile;

import org.apache.kafka.clients.producer.*;

public class PassthruProducer implements IProducer {

  private final KafkaProducer<String, String> producer;
  private final String topic;

  public PassthruProducer(String servers, String topic) {
    this.producer = new KafkaProducer(
      IProducer.createConfig(servers));          ◁────── Use the IProducer
    this.topic = topic;                                  interface's createConfig
  }                                                      function to configure
                                                         the producer.
  public void process(String message) {
    IProducer.write(this.producer, this.topic, message);  ◁── Write each
  }                                                            supplied record out
}                                                            to the specified
                                                             Kafka topic.
```

The PassthruProducer implementation should be fairly self-explanatory; it simply writes out each supplied message to a new Kafka topic.

### 3.3.5  *Stitching it all together*

All that's left is to stitch these three files together via a new StreamApp class containing our main method. Create a new file called src/main/java/nile/StreamApp.java and populate it with the contents of the following listing.

**Listing 3.5  StreamApp.java**

```
package nile;

public class StreamApp {

  public static void main(String[] args){
    String servers  = args[0];
    String groupId  = args[1];
    String inTopic  = args[2];
    String goodTopic = args[3];

    Consumer consumer = new Consumer(servers, groupId, inTopic);
    PassthruProducer producer = new PassthruProducer(
```

```
       servers, goodTopic);
    consumer.run(producer);
  }
}
```

We will pass four arguments into our `StreamApp` on the command-line:

- `servers` specifies the host and port for talking to Kafka.
- `groupId` identifies our code as belonging to a specific Kafka consumer group.
- `inTopic` is the Kafka topic we will read from.
- `goodTopic` is the Kafka topic we will write all events to.

Let's build our stream processing app now. From the project root, the nile folder, run this:

```
$ gradle jar
...
BUILD SUCCESSFUL

Total time: 25.532 secs
```

Great—we are now ready to test our stream processing app.

### 3.3.6   Testing

To test out our new application, we are going to need five terminal windows. Figure 3.8 sets out what we'll be running in each of these terminals.

**Figure 3.8   The five terminals we need to run to test our initial Kafka worker include ZooKeeper, Kafka, one topic producer, one consumer, and the app itself.**

Our first four terminal windows will each run a shell script from inside our Kafka installation directory:

```
$ cd ~/kafka_2.12-2.0.0
```

In our first terminal, we start up ZooKeeper:

```
$ bin/zookeeper-server-start.sh config/zookeeper.properties
```

In our second terminal, we start up Kafka:

```
$ bin/kafka-server-start.sh config/server.properties
```

In our third terminal, let's start a script that lets us send events into our raw-events Kafka topic. We'll call this raw-events-ch03 to prevent any clashes with our work in chapter 2:

```
$ bin/kafka-console-producer.sh --topic raw-events-ch03 \
  --broker-list localhost:9092
```

Let's now give this producer some events, by pasting these into the same terminal:

```
{ "event": "SHOPPER_VIEWED_PRODUCT", "shopper": { "id": "123",
 "name": "Jane", "ipAddress": "70.46.123.145" }, "product": { "sku":
"aapl-001", "name": "iPad" }, "timestamp": "2018-10-15T12:01:35Z" }
{ "event": "SHOPPER_VIEWED_PRODUCT", "shopper": { "id": "456",
 "name": "Mo", "ipAddress": "89.92.213.32" }, "product": { "sku":
"sony-072", "name": "Widescreen TV" }, "timestamp":
"2018-10-15T12:03:45Z" }
{ "event": "SHOPPER_VIEWED_PRODUCT", "shopper": { "id": "789",
 "name": "Justin", "ipAddress": "97.107.137.164" }, "product": {
"sku": "ms-003", "name": "XBox One" }, "timestamp":
"2018-10-15T12:05:05Z" }
```

Note that you need a newline between each event to send it into the Kafka topic. Next, in our fourth terminal, we'll start a script to "tail" our outbound Kafka topic:

```
$ bin/kafka-console-consumer.sh --topic enriched-events --from-beginning \
  --bootstrap-server localhost:9092
```

Phew! We are finally ready to start up our new stream processing application. In a fifth terminal, head back to your project root, the nile folder, and run this:

```
$ cd ~/nile
$ java -jar ./build/libs/nile-0.1.0.jar localhost:9092 ulp-ch03-3.3 \
  raw-events-ch03 enriched-events
```

This has kicked off our app, which will now read all events from raw-events-ch03 and mirror them directly to enriched-events. Check back in the fourth terminal (the console-consumer), and you should see our three events appearing in the enriched-events Kafka topic:

```
{ "event": "SHOPPER_VIEWED_PRODUCT", "shopper": { "id": "123",
 "name": "Jane", "ipAddress": "70.46.123.145" }, "product": { "sku":
"aapl-001", "name": "iPad" }, "timestamp": "2018-10-15T12:01:35Z" }
{ "event": "SHOPPER_VIEWED_PRODUCT", "shopper": { "id": "456",
 "name": "Mo", "ipAddress": "89.92.213.32" }, "product": { "sku":
"sony-072", "name": "Widescreen TV" }, "timestamp":
"2018-10-15T12:03:45Z" }
{ "event": "SHOPPER_VIEWED_PRODUCT", "shopper": { "id": "789",
 "name": "Justin", "ipAddress": "97.107.137.164" }, "product": {
"sku": "ms-003", "name": "XBox One" }, "timestamp":
"2018-10-15T12:05:05Z" }
```

Good news: our simple pass-through stream processing app is working a treat. Now we can move onto the more complex version involving event validation and enrichment.

Shut down the stream processing app with Ctrl-Z and then type kill %%, but make sure to leave that terminal and the other terminal windows open for the next section.

## 3.4    Writing a single-event processor

The next step involves developing our simple Kafka worker into a complete *single-event processor*, which handles validation, enrichment, and routing as per Nile's requirements. Let's get started.

### 3.4.1    Writing our event processor

We are in luck: because we built our pipeline in section 3.3 around a Java interface, called IProducer, we can swap out our existing PassthruProducer with a more sophisticated event processor with a minimum of fuss. Let's remind ourselves first of what our Nile bosses want this event processor to do:

- *Read* events from our Kafka topic raw-events
- *Validate* the events, writing any validation failures to the bad-events Kafka topic
- *Enrich* our validated events with the geographical location of the shopper by using the MaxMind geo-IP database
- *Write* our validated, enriched events to the enriched-events Kafka topic

In the interest of simplicity, we will use a simple definition of a *valid event*—namely, an event that does the following:

- Contains a shopper.ipAddress property, which is a string
- Allows us to add a shopper.country property, which is also a string, without throwing an exception

If these conditions are not met, we will generate an error message, again in JSON format, and write this to the bad-events topic in Kafka. Our error messages will be simple:

```
{ "error": "Something went wrong" }
```

For this section, as you prefer, you can either make a full copy of the nile codebase from section 3.3, or make changes *in situ* in that codebase. Either way, you will first need to create a file, src/main/java/nile/FullProducer.java, and paste in the contents of the following listing.

#### Listing 3.6    FullProducer.java

```java
package nile;

import com.fasterxml.jackson.databind.*;
import com.fasterxml.jackson.databind.node.ObjectNode;

import java.net.InetAddress;
import org.apache.kafka.clients.producer.*;
```

```
import com.maxmind.geoip2.*;
import com.maxmind.geoip2.model.*

public class FullProducer implements IProducer {

  private final KafkaProducer<String, String> producer;
  private final String goodTopic;
  private final String badTopic;
  private final DatabaseReader maxmind;

  protected static final ObjectMapper MAPPER = new ObjectMapper();

  public FullProducer(String servers, String goodTopic,
    String badTopic, DatabaseReader maxmind) {
    this.producer = new KafkaProducer(
      IProducer.createConfig(servers));
    this.goodTopic = goodTopic;
    this.badTopic = badTopic;
    this.maxmind = maxmind;
  }

  public void process(String message) {

    try {
      JsonNode root = MAPPER.readTree(message);
      JsonNode ipNode = root.path("shopper").path("ipAddress");
      if (ipNode.isMissingNode()) {
        IProducer.write(this.producer, this.badTopic,
          "{\"error\": \"shopper.ipAddress missing\"}");
      } else {
        InetAddress ip = InetAddress.getByName(ipNode.textValue());
        CityResponse resp = maxmind.city(ip);
        ((ObjectNode)root).with("shopper").put(
          "country", resp.getCountry().getName());
        ((ObjectNode)root).with("shopper").put(
          "city", resp.getCity().getName());
        IProducer.write(this.producer, this.goodTopic,
          MAPPER.writeValueAsString(root));
      }
    } catch (Exception e) {
      IProducer.write(this.producer, this.badTopic, "{\"error\": \"" +
        e.getClass().getSimpleName() + ": " + e.getMessage() + "\"}");
    }
  }
}
```

Annotations:

- **Constructor takes good and bad Kafka topics for writing, plus the MaxMind geo-IP lookup service**
- **Retrieves the ipAddress from the shopper object within the incoming event**
- **Looks up the shopper's location based on shopper's IP address**
- **In case of validation or processing failure, writes error message out to "bad" Kafka topic**
- **Adds the shopper's country and city to the event**
- **Writes the now-enriched event out to our "good" Kafka topic**

There's quite a lot to take in here. The control flow is perhaps better visualized in a diagram, as shown in figure 3.9. The important thing to understand is that we are looking up the shopper's IP address in MaxMind, and if it's found, we are attaching the shopper's country and city to the outgoing enriched event. If anything goes wrong on the way, we write that error message out to the "bad" topic.

As you've probably guessed, we will need to make some tweaks to our app's main function to support the new MaxMind functionality. Let's do that now.

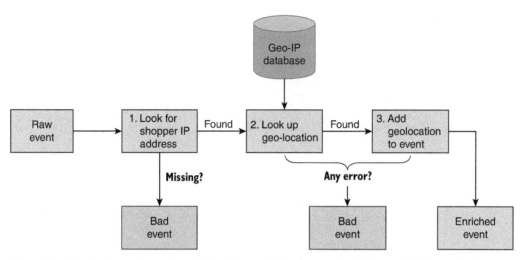

**Figure 3.9** Our single-event processor attempts to enrich the raw event with the geolocation as looked up in the MaxMind database; if anything goes wrong, an error is written out instead.

### 3.4.2 Updating our main function

Head back to src/main/java/nile/StreamApp.java and make the additions set out in the following listing.

**Listing 3.7 StreamApp.java**

```
package nile;

import java.io.*;                    ◁─── Initializing the MaxMind
                                          geo-IP database can throw
import com.maxmind.geoip2.DatabaseReader;  an exception.

public class StreamApp {

    public static void main(String[] args) throws IOException {   ┐
        String servers     = args[0];
        String groupId     = args[1];                               New
        String inTopic     = args[2];                               command-line
        String goodTopic   = args[3];                               arguments
        String badTopic    = args[4];
        String maxmindFile = args[5];                             ┘

        Consumer consumer = new Consumer(servers, groupId, inTopic);
        DatabaseReader maxmind = new DatabaseReader.Builder(new
            File(maxmindFile)).build();
        FullProducer producer = new FullProducer(          Initialize the FullProducer with
            servers, goodTopic, badTopic, maxmind);   ◁─── both outbound Kafka topics and
        consumer.run(producer);                            the MaxMind geo-IP database.
    }
}
```

Initialize the MaxMind geo-IP database.

Note the two new arguments in our `StreamApp`:

- `badTopic` is the Kafka topic we will write errors to.
- `maxmindFile` is the full path to the MaxMind geo-IP database.

Before we build the application, open the build.gradle file in the root and change

```
version = '0.1.0'
```

to

```
version = '0.2.0'
```

Let's rebuild our stream processing app now. From the project root, the nile folder, run this:

```
$ gradle jar
...
BUILD SUCCESSFUL

Total time: 25.532 secs
```

Great—now we can test this!

### 3.4.3 *Testing, redux*

Before we can run our app, we need to download a free copy of the MaxMind geo-IP database. You can do this like so:

```
$ wget \
  "https://geolite.maxmind.com/download/geoip/database/GeoLite2-City.tar.gz"
$ tar xzf GeoLite2-City_<yyyyMMdd>.tar.gz
```

To run our event processor, type in the following:

```
$ java -jar ./build/libs/nile-0.2.0.jar localhost:9092 ulp-ch03-3.4 \
  raw-events-ch03 enriched-events bad-events ./GeoLite2-
    City_<yyyyMMdd>/GeoLite2-City.mmdb
```

Great—our app is now running! Note that we configured it with a different consumer group to the previous app: `ulp-ch03-3.4` versus `ulp-ch03-3.3`. Therefore, this app will process events right back from the start of the `raw-events-ch03` topic. If you've left everything running from section 3.3, our events should be flowing through our *single-event processor* now.

Check back in the fourth terminal (the `console-consumer`) and you should see our original three raw events appearing in the `enriched-events` Kafka topic, but this time with the geolocation data attached—namely, the country and city fields:

```
{"event":"SHOPPER_VIEWED_PRODUCT","shopper":{"id":"123","name":"Jane",
 "ipAddress":"70.46.123.145","country":"United States", "city":
 "Greenville"}, "product":{"sku":"aapl-001", "name":"iPad"},
 "timestamp": "2018-10-15T12:01:35Z"}
{"event":"SHOPPER_VIEWED_PRODUCT","shopper":{"id":"456","name":"Mo",
```

```
  "ipAddress":"89.92.213.32","country":"France","city": "Rueil-malmaison"},
  "product":{"sku":"sony-072","name":"Widescreen TV"},"timestamp":
  "2018-10-15T12:03:45Z"}
{"event":"SHOPPER_VIEWED_PRODUCT","shopper":{"id":"789","name": "Justin",
  "ipAddress":"97.107.137.164","country":"United States","city":
  "Absecon"}, "product":{"sku":"ms-003","name":"XBox One"}, "timestamp":
  "2018-10-15T12:05:05Z"}
```

This looks great! We are successfully enriching our incoming events, adding useful geographical context to these events for the Nile data scientists. There's just one more thing to check—namely, that our single-event processor handles corrupt or somehow invalid events correctly. Let's send some in. Switch to back to the third console, which is running the following:

```
$ bin/kafka-console-producer.sh --topic raw-events-ch03 \
  --broker-list localhost:9092
```

Let's now feed our stream processing app corrupt events, by pasting the following into the same terminal:

```
not json
{ "event": "SHOPPER_VIEWED_PRODUCT", "shopper": { "id": "456", "name":
  "Mo", "ipAddress": "not an ip address" }, "product": { "sku": "sony-072",
  "name": "Widescreen TV" }, "timestamp": "2018-10-15T12:03:45Z" }
{ "event": "SHOPPER_VIEWED_PRODUCT", "shopper": {}, "timestamp":
  "2018-10-15T12:05:05Z" }
```

Note that you need a newline between each event to send it into the Kafka topic. To test this, we are going to need one additional terminal. This will tail the bad-events Kafka topic, which will contain our event validation failures. Let's start the consumer script like so:

```
$ bin/kafka-console-consumer.sh --topic bad-events --from-beginning \
  --bootstrap-server localhost:9092
```

By way of a sense-check, you should now have a six-pane terminal layout, as per figure 3.10.

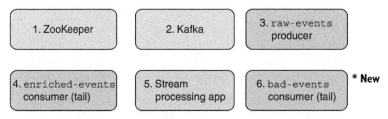

**Figure 3.10   Our six terminals consist of the same five as before, plus a second consumer, or "tail"—this time for the bad-events topic.**

Wait a few seconds and you should start to see the validation failures stream into the
bad-events topic:

```
{"error": "JsonParseException: Unrecognized token 'not':
 was expecting 'null', 'true', 'false' or NaN
 at [Source: not json; line: 1, column: 4]"}
{"error": "NullPointerException: null"}
{"error": "shopper.ipAddress missing"}
```

This completes our testing. We now have our single-event processor successfully validating the incoming events, enriching them, and routing the output to the appropriate channel.

## Summary

- We set up a unified log like Kafka and feed events into it so that we can process
  those event streams.
- Event stream processing can include backing up the stream, monitoring it,
  loading it into a database, or creating aggregates.
- Processing single events is much less complex than processing batches or windows of multiple events at a time.
- When stream processing, we can write our results out to another stream, so our
  unified log acts as the "superglue" between our company's systems.
- We created a simple Java app for Nile that ran as a "Kafka worker," reading
  events from one Kafka topic and passing them through to another Kafka topic
  unchanged.
- We extended this Java app into a single-event processor, which validated the
  incoming events, attempted to enrich them with a geographical location, and
  then wrote the output to either a "good" or "bad" Kafka topic as appropriate.
- Geographical location is a great example of information about a raw event that
  can be derived from the raw event and attached as additional context.

# Event stream processing
## with Amazon Kinesis

**This chapter covers**

- Amazon Kinesis, a fully managed unified log service
- Systems monitoring as a unified log use case
- Using the AWS CLI tools to work with Kinesis
- Building simple Kinesis producers and consumers in Python

So far in this book, we have worked exclusively with Apache Kafka as our unified log. Because it is an open source technology, we have had to set up and configure Kafka and its dependencies (such as ZooKeeper) ourselves. This has given us great insight into how a unified log works "under the hood," but some of you may be wondering whether there is an alternative that is not so operationally demanding. Can we outsource the operation of our unified log to a third party, without losing the great qualities of the unified log?

The answer is a qualified yes. This chapter introduces Amazon Kinesis (https://aws.amazon.com/kinesis/), a hosted unified log service available as part of Amazon Web Services. Developed internally at Amazon to solve its own challenges around log collection at scale, Kinesis has extremely similar semantics to Kafka—along with subtle differences that we will tease out in this chapter.

Before kicking off this chapter, you might want to jump to the appendix for a brief AWS primer that will get you up to speed on the Amazon Web Services platform, unless you already know your way into the AWS ecosystem. Once your AWS account is set up, this chapter will show you how to use the AWS command-line interface (CLI) tools to create your first event stream in Kinesis and write and read some events to it.

We will then dive into a new use case for the unified log: using it for *systems monitoring*. We will create a simple long-running *agent* in Python that emits a steady stream of readings from our server, writing these events to Kinesis by using the AWS Python SDK. Once we have these events in Kinesis, we will write another Python application that *monitors* our agent's events, looking for potential problems. Again, this Python monitoring application will be built using the AWS Python SDK, also known as *boto*.

Please note that, as Amazon Kinesis Data Streams is not currently available in AWS Free Tier, the procedures in this book will necessarily involve creating live resources in your Amazon Web Services account, which can incur some charges.[1] Don't worry— I will tell you as soon as you can safely delete a given resource. In addition, you can set alerts on your spending in order to be notified whenever the charges go above a certain threshold.[2]

## 4.1 *Writing events to Kinesis*

Our AWS account is set up, and we have the AWS CLI primed for action: we're ready to introduce the project we'll be working through! In previous chapters, we focused on events related to *end-user behavior*. In this chapter, we'll take a different tack and generate a simple stream of events related to *systems monitoring*.

### 4.1.1 *Systems monitoring and the unified log*

Let's imagine that our company has a server that keeps running out of space. It hosts a particularly chatty application that keeps generating lots of log files. Our systems administrator wants to receive a warning whenever the server's disk reaches 80% full, so that he can go in and manually archive and remove the excess log files.

Of course, there is a rich, mature ecosystem of systems monitoring tools that could meet this requirement. Simplifying somewhat,[3] these tools typically use one of two monitoring architectures:

- *Push-based monitoring*—An agent running on each monitored system periodically sends data (often called *metrics*) into a centralized system. Push-based monitoring systems include Ganglia, Graphite, collectd, and StatsD.

---

[1] You can find more information about the pricing of AWS Kinesis Data Streams at https://aws.amazon .com/kinesis/streams/pricing/.

[2] You can set up AWS billing notification alerts at https://docs.aws.amazon.com/awsaccountbilling/latest/ aboutv2/billing-getting-started.html#d0e1069.

[3] A good article about push versus pull monitoring is at https://web.archive.org/web/20161004192212/ https://www.boxever.com/push-vs-pull-for-monitoring.

- *Pull-based monitoring*—The centralized system periodically "scrapes" metrics from each monitored system. Pull-based monitoring systems include JMX, librit, and WMI.

Figure 4.1 depicts both the push and pull approaches. Sometimes a systems monitoring tool will provide both approaches. For example, Zabbix and Prometheus are predominantly pull-based systems with some push support.

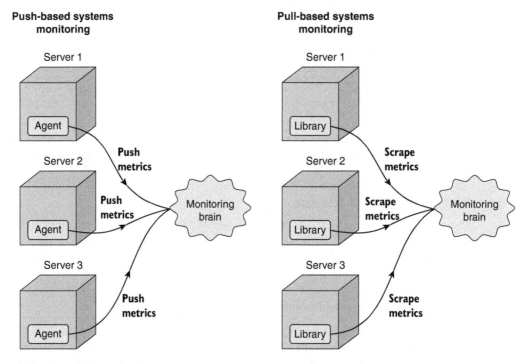

**Figure 4.1   In push-based systems monitoring, an agent pushes metrics at regular intervals into a centralized system. By contrast, in pull-based architectures, the centralized system regularly scrapes metrics from endpoints available on the servers.**

In the push-based model, we have agents generating events and submitting them to a centralized system; the centralized system then analyzes the event stream obtained from all agents. Does this sound familiar? We can transplant this approach directly into our unified log, as per figure 4.2. The building blocks are the same, as you saw in earlier chapters: event producers, a unified log, and a set of event consumers. But there are two main differences from our previous unified log experiences:

- Instead of adding event tracking to an existing application like HelloCalculator, we will be creating a dedicated *agent* that exists only to send events to our unified log.
- Rather than the *subject* of our events being an end user, the subject will now be our agent, because it is this agent that is actively taking readings from the server.

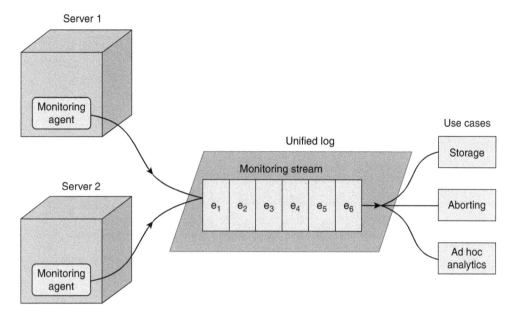

**Figure 4.2   We can implement push-based systems monitoring on top of our unified log. The agents running on our servers will emit events that are written to our unified log, ready for further processing.**

It looks, then, like it should be straightforward to meet our systems administrator's monitoring requirements by using a unified log such as Apache Kafka or Amazon Kinesis. We will be using Kinesis for this chapter, so before we get started, we'll take a brief look at the terminology differences between Kafka and Kinesis.

### 4.1.2   Terminology differences from Kafka

Amazon Kinesis has extremely similar *semantics* to Apache Kafka. But the two platforms diverge a little in the descriptive language that they use. Figure 4.3 sets out the key differences: essentially, Kinesis uses *streams*, whereas Kafka uses *topics*. Kinesis streams consist of one or more *shards*, whereas Kafka topics contain *partitions*. Personally, I prefer the Kinesis terms to Kafka's: they are a little less ambiguous and have less "message queue" baggage.

Differences of language aside, the fact that Kinesis offers the same key building blocks as Kafka is encouraging: it suggests that almost everything we could do in Kafka, we can do in Kinesis. To be sure, we will come across differences of approach and capability through this chapter; I will make sure to highlight these as they come up. For now, let's get started with Kinesis.

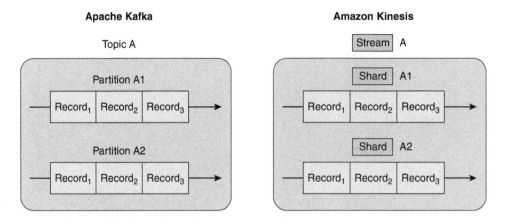

**Figure 4.3   The equivalent of a Kafka topic in Kinesis is a stream. A stream consists of one or more shards, whereas Kafka refers to partitions.**

### 4.1.3   *Setting up our stream*

First, we need a Kinesis stream to send our systems monitoring events to. Most commands in the AWS CLI follow this format:

```
$ aws [service] [command] options...
```

In our case, all of our commands will start with aws kinesis. You can find a full reference of all available AWS CLI commands for Kinesis here:

```
https://docs.aws.amazon.com/cli/latest/reference/kinesis/index.html
```

We can create our new stream by using the AWS CLI like so:

```
$ aws kinesis create-stream --stream-name events \
  --shard-count 2 --profile ulp
```

Press Enter, and then switch back to the AWS web interface, and click Amazon Kinesis. If you are quick enough, you should see the new stream listed with its status set to CREATING, and with a count of 0 shards, as in figure 4.4. After the stream is created, Kinesis will report the stream status as ACTIVE and display the correct number of shards. We can write events to and read events from only ACTIVE streams.

We created our stream with two shards; events that are sent to the event stream will be written to one or either of the two shards. Any stream processing apps that we write will have to make sure to read events from *all* shards. At the time of writing, Amazon enforces a few limits around shards:[4]

---

[4]   AWS Kinesis Data Streams limits are described at https://docs.aws.amazon.com/streams/latest/dev/service-sizes-and-limits.html.

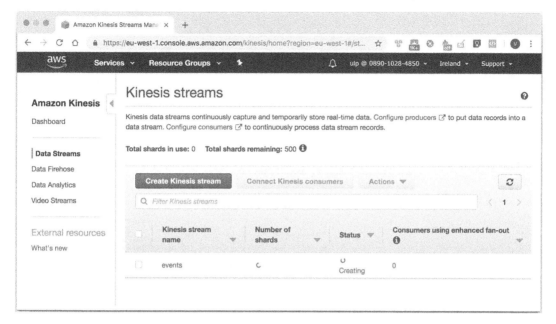

**Figure 4.4  Our first Amazon Kinesis stream is being created. After a few more seconds, we will see a status of ACTIVE and the correct shard count.**

- You are allowed 200 shards per AWS region by default, except for the following AWS regions, which allow up to 500 shards: US East (N. Virginia), US West (Oregon), and EU (Ireland).
- Each shard supports up to five read transactions per second. Each transaction can provide up to 10,000 records, with an upper limit of 10 MB per transaction.
- Each shard supports writing up to 1 MB of record data per second, and up to 1,000 records per second.

Don't worry—we won't be hitting any of these limits; we could have happily made do with just one shard in our stream.

So, at this point, we have our Kinesis stream ready and waiting to receive events. In the next section, let's model those events.

### 4.1.4   Modeling our events

Remember that our systems administrator wants to receive a warning whenever the troublesome server's disk reaches 80% full. To support this monitoring, the agent running on the server will need to regularly read filesystem metrics and send those metrics into our unified log for further analysis. We can model these metrics readings as events by using the grammatical structure introduced in chapter 2:

- *Our agent* is the subject of the event.
- *Read* ("took a reading") is the verb of the event.

- *Filesystem metrics* are the direct object of the event.
- The reading takes place on our *server*, a prepositional object.
- The reading takes place at a *specific time*, another prepositional object.

Putting these together, we can sketch out the event model that we'll need to assemble, as in figure 4.5.

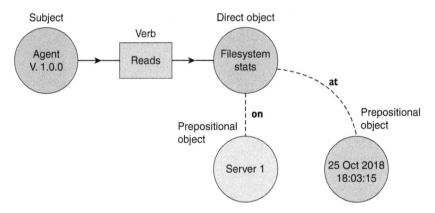

**Figure 4.5   Our systems monitoring events involve an agent reading filesystem metrics on a given server at a specific point in time.**

Now that you know what our events should look like, let's write our agent.

### 4.1.5   *Writing our agent*

We are going to write our agent in Python, making use of the excellent boto3 library, which is the official Python SDK for AWS.[5] All the official language-specific SDKs for AWS support writing events to Kinesis, which seems fair: for Kinesis to be a truly unified log, we need to be able to send events to it from all our various client applications, whatever language they are written in.

Let's get started. We are going to build up our systems monitoring agent piece by piece, using the Python interactive interpreter. Start it up by logging into your Vagrant guest and typing `python` at the command prompt:

```
Python 2.7.12 (default, Dec  4 2017, 14:50:18)
[GCC 5.4.0 20160609] on linux2
Type "help", "copyright", "credits" or "license" for more information.
>>>
```

First let's define and test a function that gives us the filesystem metrics that we need. Paste the following into your Python interpreter, being careful to keep the white-space intact:

---

[5]  You can download the Python SDK for AWS from https://aws.amazon.com/sdk-for-python/.

```
import os
def get_filesystem_metrics(path):
  stats = os.statvfs(path)
  block_size = stats.f_frsize
  return (block_size * stats.f_blocks, # Filesystem size in bytes
    block_size * stats.f_bfree,        # Free bytes
    block_size * stats.f_bavail)       # Free bytes excl. reserved space

s, f, a = get_filesystem_metrics("/")
print "size: {}, free: {}, available: {}".format(s, f, a)
```

You should see something like the following output; the exact numbers will depend on your computer:

```
size: 499046809600, free: 104127823872, available: 103865679872
```

Good—now you know how to retrieve the information we need about the filesystem. Next, let's create all the metadata that we need for our event. Paste in the following:

```
import datetime, socket, uuid
def get_agent_version():
  return "0.1.0"

def get_hostname():
  return socket.gethostname()

def get_event_time():
  return datetime.datetime.now().isoformat()

def get_event_id():
  return str(uuid.uuid4())

print "agent: {}, hostname: {}, time: {}, id: {}".format(
  get_agent_version(), get_hostname(), get_event_time(), get_event_id())
```

You should now see something a little like this:

```
agent: 0.1.0, hostname: Alexanders-MacBook-Pro.local, time:
2018-11-01T09:00:34.515459, id: 42432ebe-40a5-4407-a066-a1361fc31319
```

Note that we are uniquely identifying each event by attaching a freshly minted version 4 *UUID* as its event ID.[6] We will explore event IDs in much more detail in chapter 10.

Let's put this all together with a function that creates our event as a Python dictionary. Type in the following at the interpreter:

```
def create_event():
  size, free, avail = get_filesystem_metrics("/")
  event_id = get_event_id()
  return (event_id, {
```

---

[6] Wikipedia provides a good description of universally unique identifiers: https://en.wikipedia.org/wiki/Universally_unique_identifier.

```
      "id": event_id,
      "subject": {
        "agent": {
          "version": get_agent_version()
        }
      },
      "verb": "read",
      "direct_object": {
        "filesystem_metrics": {
          "size": size,
          "free": free,
          "available": avail
        }
      },
      "at": get_event_time(),
      "on": {
        "server": {
          "hostname": get_hostname()
        }
      }
    })

print create_event()
```

It's a little verbose, but the intent should be clear from the Python interpreter's output, which should be something like this:

```
('60f4ead5-8a1f-41f5-8e6a-805bbdd1d3f2', {'on': {'server': {'hostname':
 'ulp'}}, 'direct_object': {'filesystem_metrics': {'available':
37267378176, 'free': 39044952064, 'size': 42241163264}}, 'verb':
'read', 'at': '2018-11-01T09:02:31.675773', 'id':
'60f4ead5-8a1f-41f5-8e6a-805bbdd1d3f2', 'subject': {'agent':
{'version': '0.1.0'}}})
```

We have now constructed our first well-structured systems monitoring event! How do we send it to our Kinesis stream? It should be as simple as this:

```
def write_event(conn, stream_name):
  event_id, event_payload = create_event()
  event_json = json.dumps(event_payload)
  conn.put_record(StreamName=stream_name, Data=event_json,
PartitionKey=event_id)
```

The key method to understand here is conn.put_record, which takes three required arguments:

- The name of the stream to write to
- The data (sometimes called *body* or *payload*) of the event. We are sending this data as a Python string containing our JSON.

- The partition key for the event. This determines which shard the event is written to.

Now we just need to connect to Kinesis and try writing an event. This is as simple as the following:

```
import boto3

session = boto3.Session(profile_name="ulp")
Conn = session.client("kinesis", region_name="eu-west-1")

write_event(conn, "events")
```

Part of the reason that this code is so simple is that the AWS CLI tool that you configured earlier uses boto, the AWS SDK for Python, under the hood. Therefore, boto can access the AWS credentials that you set up earlier in the AWS CLI without any trouble.

Press Enter on the preceding code and you should be greeted with ... silence! Although, in this case, no news is good news, it would still be nice to get some visual feedback. This can be arranged: next, put the sending of our event into an infinite loop in the Python interpreter, like so:

```
while True:
  write_event(conn, "events")
```

Leave this running for a couple of minutes, and then head back into the Kinesis section of the AWS web interface, and click your events stream to bring up the Stream Details view. At the bottom of this view, in the Monitoring tab, you should be able to see the beginnings of lines on some of the charts, as in figure 4.6.

Unfortunately, this is the only visual confirmation we can get that we are successfully writing to our Kinesis stream—at least until we write some kind of stream consumer. We will do that soon, but first let's wrap up our systems monitoring agent. We won't need the Python interpreter anymore, so you can kill the infinite loop with Ctrl-C, and then exit the interpreter with Ctrl-D.

Let's consolidate all of our work at the interpreter into a single file to run our agent's core monitoring loop. Create a file called agent.py and populate it with the contents of the following listing.

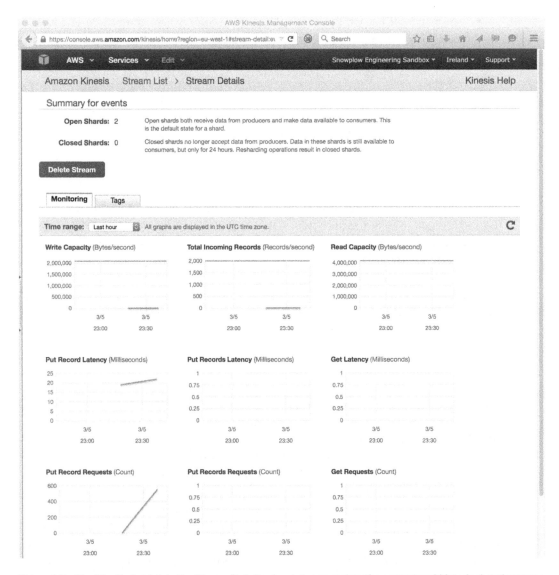

**Figure 4.6    The Monitoring tab in the Stream Details view lets you review the current and historical performance of a given Kinesis stream.**

Listing 4.1    agent.py

```python
#!/usr/bin/env python

import os, datetime, socket, json, uuid, time, boto3

def get_filesystem_metrics(path):
  stats = os.statvfs(path)
  block_size = stats.f_frsize
```

```
     return (block_size * stats.f_blocks, # Filesystem size in bytes
        block_size * stats.f_bfree,        # Free bytes
        block_size * stats.f_bavail)       # Free bytes excluding reserved space

def get_agent_version():
  return "0.1.0"

def get_hostname():
  return socket.gethostname()

def get_event_time():
  return datetime.datetime.now().isoformat()

def get_event_id():
  return str(uuid.uuid4())

def create_event():
  size, free, avail = get_filesystem_metrics("/")
  event_id = get_event_id()
  return (event_id, {
    "id": event_id,
    "subject": {
      "agent": {
        "version": get_agent_version()
      }
    },
    "verb": "read",
    "direct_object": {
      "filesystem_metrics": {
        "size": size,
        "free": free,
        "available": avail
      }
    },
    "at": get_event_time(),
    "on": {
      "server": {
        "hostname": get_hostname()
      }
    }
  })

def write_event(conn, stream_name):
  event_id, event_payload = create_event()
  event_json = json.dumps(event_payload)
  conn.put_record(StreamName=stream_name, Data=event_json,
PartitionKey=event_id)
  return event_id

if __name__ == '__main__':                    ◄─────┘ The entry point
                                                       for our app
  session = boto3.Session(profile_name="ulp")
  conn = session.client("kinesis", region_name="eu-west-1")
  while True:                                 ◄─────┘ Loop forever.
    event_id = write_event(conn, "events")
    print (f'Wrote event: {event_id}')         Emit one event
    time.sleep(10)                      ◄─────┘ every 10 seconds.
```

Make the agent.py file executable and run it:

```
chmod +x agent.py
./agent.py
Wrote event 481d142d-60f1-4d68-9bd6-d69eec5ff6c0
Wrote event 3486558d-163d-4e42-8d6f-c0fb91a9e7ec
Wrote event c3cd28b8-9ddc-4505-a1ce-193514c28b57
Wrote event f055a8bb-290c-4258-90f0-9ad3a817b26b
...
```

Our agent is running! Check out figure 4.7 for a visualization of what we have created.

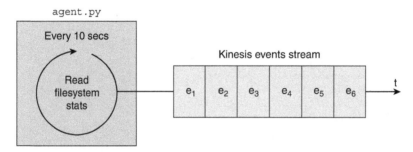

**Figure 4.7  Our systems monitoring agent is now emitting an event containing filesystem statistics every 10 seconds.**

Leave our Python agent running in a terminal. It will continue to write an *Agent read filesystem metrics* event every 10 seconds. Now we need to write our stream processing app to consume this stream of events and notify us if our server's disk reaches the dreaded 80% full.

## 4.2    Reading from Kinesis

A variety of frameworks and SDKs can read events from a Kinesis stream. Let's review these briefly, and then use two of these tools to implement basic monitoring of our event stream.

### 4.2.1    Kinesis frameworks and SDKs

For such a young platform, Kinesis already supports a slightly bewildering array of frameworks and SDKs for event stream processing. Here are the main ones:

- The AWS CLI introduced earlier has a small set of commands to help with reading records from a given shard within a Kinesis stream.[7]
- Each official AWS SDK has support for reading records from a Kinesis stream. Your calling application is expected to run a thread for each shard within the

---

[7]  See the AWS CLI Command Reference for a list of these commands: https://docs.amazon.com/cli/latest/reference/kinesis/index.html#cli-aws-kinesis.

stream, and you are responsible for keeping track of your current processing position within each shard.

- A higher-level framework, again written by the AWS team, is called the Kinesis Client Library (KCL) for Java.[8] This uses the Amazon DynamoDB database to keep track of your calling application's processing positions within each shard. It also handles the "division of labor" between multiple instances of a KCL-powered application running on separate servers, which is useful for horizontal scaling.

- The KCL for Java includes a `MultiLangDaemon`, which enables Kinesis Client Library applications to be written in other languages. At the time of writing, there is only a Kinesis Client Library for Python built using this.[9]

- AWS Lambda is a fully managed stream processing platform running on a Node.js cluster. You write and upload JavaScript functions and assign them to be run for every event in an AWS-native event stream such as Kinesis. Functions must be short-lived (completing in no more than 15 minutes) and cannot use local state. It is also possible to write functions in Java, Python, Go, Ruby, and C#.

- Apache Storm has a Kinesis Storm Spout, created by the AWS team, which retrieves data records from Amazon Kinesis and emits them as tuples, ready for processing in Storm.

- Apache Spark Streaming, which is Spark's microbatch processing framework, can convert a Kinesis stream into an `InputDStream` ready for further processing. This functionality is built on top of the KCL for Java.

A surprising omission from the preceding list is Apache Samza—surprising because, as you have seen, the semantics of Amazon Kinesis and Apache Kafka are extremely similar. Given how well Samza works with Kafka (see chapter 5), we might expect it to work well with Kinesis too.

### 4.2.2 *Reading events with the AWS CLI*

Let's start with the "rawest" tool for processing a Kinesis stream: the AWS CLI. You would likely choose a higher-level framework to build a production application, but the AWS CLI will give you familiarity with the key building blocks underpinning those frameworks.

First, we'll use the `describe-stream` command to review the exact contents of our Kinesis stream:

```
$ aws kinesis describe-stream --stream-name events --profile ulp
{
    "StreamDescription": {
        "Shards": [
```

---

[8] You can download the official Kinesis Client library for Java from GitHub at https://github.com/awslabs/amazon-kinesis-client.

[9] You can download the official Kinesis Client library for Python from https://github.com/awslabs/amazon-kinesis-client-python.

```
        {
            "ShardId": "shardId-000000000000",
            "HashKeyRange": {
                "StartingHashKey": "0",
                "EndingHashKey": "170141183460469231731687303715884105727"
            },
            "SequenceNumberRange": {
                "StartingSequenceNumber":
    "49589726466290061031074327390112813890652759903239667714"
            }
        },
        {
            "ShardId": "shardId-000000000001",
            "HashKeyRange": {
                "StartingHashKey": "170141183460469231731687303715884105728",
                "EndingHashKey": "340282366920938463463374607431768211455"
            },
            "SequenceNumberRange": {
                "StartingSequenceNumber":
    "49589726466312361776272858013254349608925408264745648146"
            }
        }
    ],
    "StreamARN": "arn:aws:kinesis:eu-west-1:089010284850:stream/events",
    "StreamName": "events",
    "StreamStatus": "ACTIVE",
    "RetentionPeriodHours": 24,
    "EnhancedMonitoring": [
        {
            "ShardLevelMetrics": []
        }
    ],
    "EncryptionType": "NONE",
    "KeyId": null,
    "StreamCreationTimestamp": 1541061618.0
    }
}
```

The response is a JSON structure containing an array of Shards, which in turn contains definitions for our stream's two shards. Each shard is identified by a unique ShardId, and contains metadata:

- The HashKeyRange, which determines which events will end up in which partition, depending on the hashed value of the event's partition key. This is illustrated in figure 4.8.
- The SequenceNumberRange, which records the sequence number of the first event in the stream. You will see only an upper bound on this range if the shard has been closed and no further events will be added.

Before we can read events from our stream, we need to retrieve what Amazon calls a *shard iterator* for each shard in the stream. A shard iterator is a slightly abstract concept. You can think of it as something like a short-lived (five-minute) file handle on a

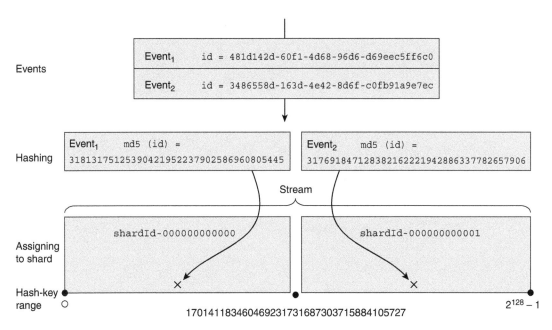

**Figure 4.8** By applying an MD5 hash to our shard key (which is our event ID), we can determine which shard Kinesis will store the event in, based on the shards' individual hash key ranges.

given shard. Each shard iterator defines a cursor position for a set of sequential reads from the shard, where the cursor position takes the form of the sequence number of the first event to read. Let's create a shard iterator now:

```
$ aws kinesis get-shard-iterator --stream-name=events \
  --shard-id=shardId-000000000000 --shard-iterator-type=TRIM_HORIZON \
  --profile=ulp
{
    "ShardIterator": "AAAAAAAAAAFVbPjgjXyjJOsE5r4/MmA8rntidIRFxTSs8rKLXSs8
kfyqcz2KxyHs3V9Ch4WFWVQvzj+xO1yWZ1rNWNjn7a5R3u0aGkMj11U2pemcJHfjkDmQKcQDwB
1qbjTdN1DzRLmYuI3u1yNDIfbG+veKBRLlodMkZOqnMEOY3bJhluDaFlOKUrynTnZ3oNA2/4zE
7uE="
}
```

We specified that the shard iterator should be of type TRIM_HORIZON. This is AWS jargon for the oldest events in the shard that have not yet been *trimmed*—expired for being too old. At the time of writing, records are trimmed from a Kinesis stream after a fixed period of 24 hours. This period can be increased up to 168 hours but will incur an additional cost. There are three other shard iterator types:

- LATEST—This returns the "most recent" data from the shard.
- AT_SEQUENCE_NUMBER—This lets us read a shard starting from the event with the specified sequence number.
- AFTER_SEQUENCE_NUMBER—This lets us read a shard starting from the event immediately after the one with the specified sequence number.

Figure 4.9 illustrates the various shard iterator options.

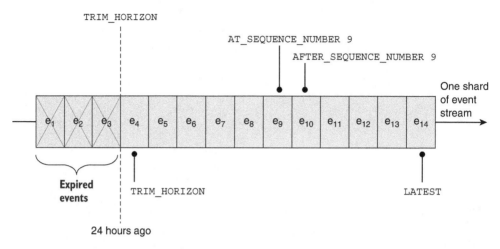

**Figure 4.9  The four configuration options for a Kinesis shard iterator allow us to start reading a shard of our event stream from various points.**

Now we are ready to try reading some records by using our new shard iterator:

```
$ aws kinesis get-records --shard-iterator
    "AAAAAAAAAAFVbPjgjXyjJOsE5r4/MmA8rntidIRFxTSs8rKLXSs8kfyqcz2KxyHs3V9Ch4WFW
VQvzj+xO1yWZ1rNWNjn7a5R3u0aGkMjl1U2pemcJHfjkDmQKcQDwB1qbjTdN1DzRLmYuI3u1yN
DIfbG+veKBRLlodMkZOqnMEOY3bJhluDaFlOKUrynTnZ3oNA2/4zE7uE=" --profile=ulp
{
    "Records": [],
    "NextShardIterator": "AAAAAAAAAAHQ8eRw4sduIDKhasXSpZtpkI4/uMBsZ1+ZrgT8
/Xg0KQ5GwUqFMIf9ooaUicRpfDVfqWRMUQ4rzYAtDIHuxdJSeMcBYX0RbBeqvc2AIRJH6BOXC6
nqZm9qJBGFIYvqb7QUAWhEFz56cnO/HLWAF1x+HUd/xT21iE3dgAFszY5H5aInXJCw+vfid4Yn
O9PZpCU="
}
```

An empty array of records! That's okay; this simply means that this initial portion of the shard doesn't contain any events. The AWS documentation warns us that we might have to cycle through a few shard iterators before we reach any events. Fortunately, the NextShardIterator gives us the next "file handle" to use, so let's plug this into our next get-records call:

```
$ aws kinesis get-records --shard-iterator
"AAAAAAAAAAEXsqVd9FvzqV7/M6+Dbz989dpSBkaAbn6/cESUTbKHNejQ3C3BmjKfRR57jQuQb
Vhlh+uN6xCOdJ+KIruWvqoITKQk9JsHa96VzJVGuLMY8sPy8Rh/LGfNSRmKO7CkyaMSbEqGNDi
gtjz7q0S41O4KL5BFHeOvGce6bJK7SJRA4BPXBITh2S1rGI62N4z9qnw=" --profile=ulp
{
    "Records": [
        {
            "PartitionKey": "b5ed136d-c879-4f2e-a9ce-a43313ce13c6",
            "Data": "eyJvbiI6IHsic2VydmVyIjogeyJob3N0bmFtZSI6ICJ1bHAifX0sI
```

CJkaXJlY3Rfb2JqZWN0IjogeyJmaWxlc3lzdGVtX21ldHJpY3MiOiB7ImF2YWlsYWJsZSI6IDM
3MjY2OTY4NTc2LCAiZnJlZSI6IDM5MDQ0NTQyNDY0LCAic2l6ZSI6IDQyMjQxMTYzMjY0fX0sI
CJ2ZXJiIjogInJlYWQiLCAiYXQiOiAiMjAxOC0wMy0wMFQyMjo1MTo1Ny4wNjYzMzUiLCAiaWQ
IOiAiYjVlZDEzNmQtYzg3OS00ZjJlLWE5Y2UtYTQzMzEzY2UxM2M2IiwgInN1YmplY3QiOiB7I
mFnZW50IjogeyJ2ZXJzaW9uIjogIjAuMS4wIn19fQ==",
            "SequenceNumber": "495485258605876791722332484369328535405053 9
8606492073986"
        },
        ...
    ],
    "NextShardIterator":"AAAAAAAAAHBdaV/lN3TN2LcaXhd9yYb45IPOc8mR/ceD5vpw
uUG0Ql5pj9UsjlXikidqP4J9HUrgGaliPLNGm+DoTH0Y8zitlf9ryiBNueeCMmhZQ6jX22yani
YKz4nbxDTKcBXga5CYDPpmj9Xb9k9A4d53bIMmIPF8JATorzwgoEilw/rbiK1a6XRdb0vDj5VH
fwzSYQ="
}

I have elided the output, but this time the AWS CLI returns 24 records, along with a `NextShardIterator` for us to fetch further events from our shard. Let's just check that the Base64-encoded contents of an event are as we expect. Again, in the Python interpreter, type in the following:

```
import base64
base64.b64decode("eyJvbiI6IHsic2VydmVyIjogeyJob3N0bmFtZSI6ICJ1bHAifX0sICJk
AXJlY3Rfb2JqZWN0IjogeyJmaWxlc3lzdGVtX21ldHJpY3MiOiB7ImF2YWlsYWJsZSI6IDM3Mj
Y2OTY4NTc2LCAiZnJlZSI6IDM5MDQ0NTQyNDY0LCAic2l6ZSI6IDQyMjQxMTYzMjY0fX0sICJ2
ZXJiIjogInJlYWQiLCAiYXQiOiAiMjAxOC0wMy0wMFQyMjo1MTo1Ny4wNjYzMzUiLCAiaWQiOi
AiYjVlZDEzNmQtYzg3OS00ZjJlLWE5Y2UtYTQzMzEzY2UxM2M2IiwgInN1YmplY3QiOiB7ImFn
ZW50IjogeyJ2ZXJzaW9uIjogIjAuMS4wIn19fQ==")
```

And you should see this:

```
{"on": {"server": {"hostname": "ulp"}}, "direct_object":
    {"filesystem_metrics": {"available": 37266968576, "free":
 39044542464, "size": 42241163264}}, "verb": "read", "at":
 "2018-11-01T09:02:31.675773", "id":
 "b5ed136d-c879-4f2e-a9ce-a43313ce13c6", "subject": {"agent":
 {"version": "0.1.0"}}}'
```

Good—we can finally confirm that our systems monitoring agent has faithfully recorded our event contents in Kinesis using boto.

Another thing to stress is that, just as in Kafka, each of these records is still stored in the Kinesis stream, available for other applications to consume. It's *not* like the act of reading has "popped" these events off the shard forever. We can demonstrate this quickly by creating an all-new shard iterator, this time laser-focused on this same event:

```
$ aws kinesis get-shard-iterator --stream-name=events \
  --shard-id=shardId-000000000000 \
  --shard-iterator-type=AT_SEQUENCE_NUMBER \
  --starting-sequence-
    number=4954852586058767917223324843693285354050539 8606492073986 \
  --profile=ulp
```

```
{
    "ShardIterator":"AAAAAAAAAAE+WN9BdSD2AoDrKCJBjX7buEixAm6FdEkHHMTYl3MgrpsmU
UOp8Q0/yd0x5zPombuawVhr6t/14zsavYqpXo8PGlex6bkvvGhRYLVeP1BxUfP91JVJicfpKQP
3Drxf0dxYeTfw6izIMUN6QCvxEluR6Ca3t0INFzpvXDIm6y36EIGpxrYmxUD0fgXbHPRdL/s="
}
```

We then request a single event (`--limit=1`) from our shard by using the new shard iterator:

```
$ aws kinesis get-records --limit=1 --shard-iterator
    "AAAAAAAAAAE+WN9BdSD2AoDrKCJBjX7buEixAm6FdEkHHMTYl3MgrpsmUUOp8Q0/yd0x5zP
    om
BuawVhr6t/14zsavYqpXo8PGlex6bkvvGhRYLVeP1BxUfP91JVJicfpKQP3Drxf0dxYeTfw6iz
IMUN6QCvxEluR6Ca3t0INFzpvXDIm6y36EIGpxrYmxUD0fgXbHPRdL/s=" --profile=ulp
{
    "Records": [
        {
            "PartitionKey": "b5ed136d-c879-4f2e-a9ce-a43313ce13c6",
            "Data":"eyJvbiI6IHsic2VydmVyIjogeyJob3N0bmFtZSI6ICJlbHAifX0sIC
JkaXJlY3Rfb2JqZWN0IjogeyJmaWWxlc3lzdGVtTX2l1dHJpY3MiOiB7ImF2YWlsYWJsZSI6IDM3
MjY2OTY4NTc2LCAiZnJlZSI6IDM5MDQ0NTQyNDY5LCAic2l6ZSI6IDQyMjQxMTYzMjY0X0sIC
J2ZXJJiIjogInJlYWQiLCAiYXQiOiAiMjAxNS0wMy0xMFQyMjo1MTo4wNjY3MzUiLCAiaWQi
OiAiYjVlZDEzNmQtYzg3OS00ZjJlLWE5Y2UtYTQzMzEzY2UxM2IiwgInN1YmplY3QiOiB7Im
FnZW50IjogeyJ2ZXXJzaW9uIjogIjAuMS4wIn19fQ==",
            "SequenceNumber":
    "49548525860587679172233248436932853540505398606492073986"
        }
    ],
    "NextShardIterator":"AAAAAAAAAFqCzzLKNkxsGFGhqUlmMHTXq/Z/xsIDu6gP+LVd
4s+KZtiPSib0mqXRiNPSEyshvmdHrV4bEwYPvxNYKLIr3xCH4T3IeSS9hdGiQsLgjJQ1yTUTe+
0qg+UJSzba/xRB7AtQURMj0xZe3sCSEjas3pzhw48uDSLyQsZu5ewqcBLja50ykJkXHOmGnCXI
oxtYMs="
}
```

This is the same `PartitionKey` (our event ID), `SequenceNumber`, and indeed Base64-encoded data as before; we have successfully retrieved the same event twice!

No doubt, this section has been a lot to take in. Let's summarize before we move on:

- The AWS CLI lets us read the events in a Kinesis stream.
- We require a *shard iterator* to read events from a single shard in the stream. Think of this as a temporary stream handle defining our cursor position in the stream.
- We use the shard iterator to read a batch of events from the shard.
- Along with the batch of events, we receive back the next shard iterator, which we use to read the next batch.

Figure 4.10 illustrates this process.

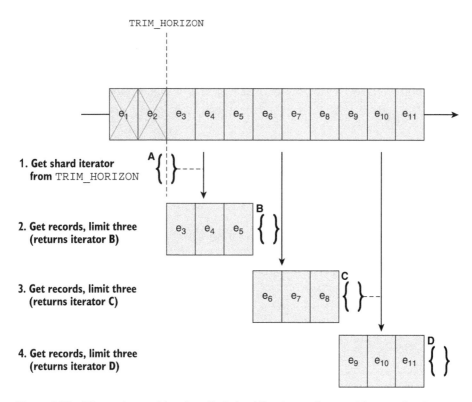

**Figure 4.10** After we have retrieved our first shard iterator, each request for records returns a new shard iterator we can use for the next request.

### 4.2.3 *Monitoring our stream with boto*

Now that you understand the basics of reading events from a Kinesis stream, let's return to the task at hand: monitoring our agent's event stream in order to check whether our server is running low on disk space. After getting our hands dirty with the AWS CLI, we're now going back to the AWS Python SDK, boto. The AWS CLI and boto expose the exact same primitives for stream processing, which is unsurprising, given that the AWS CLI is built on boto! The main difference is that the AWS Python SDK will let us use all the power of Python in our stream processing application.

The stream processing application we'll build in this section is illustrated in figure 4.11. It follows a simple algorithm:

- Read each event from each shard of our `events` stream
- Check the event's filesystem metrics to see whether our server's disk is more than 80% full
- If the disk is more than 80% full, print out an alert to the console

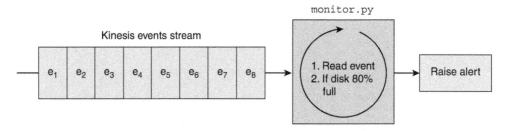

**Figure 4.11  Our monitoring application will read events from our Kinesis stream, check whether the reported disk usage has reached 80%, and raise an alert if so.**

As before, we are going to build up our application piece by piece in the Python interpreter. This time, however, we'll start with the stream reading mechanics and then work our way back to the business logic (the event monitoring).

First, we need to create a thread for each of the shards in our event stream. Although we know that our stream has two shards, this could change in the future, so we'll use boto to check the number of shards our stream currently has. Type the following at your Python prompt:

```
mport boto3

session = boto3.Session(profile_name="ulp")
conn = session.client("kinesis", region_name="eu-west-1")

stream = conn.describe_stream(StreamName='events')
shards = stream['StreamDescription']['Shards']
print (f'Shard count: {len(shards)}')
```

Press Enter and you should see this:

```
Shard count: 2
```

This is all the information we need to create a thread for each shard:

```
from threading import Thread, current_thread

def ready(shard_id):
  name = current_thread().name
  print(f'{name} ready to process shard {shard_id}')

for shard_idx in range(len(shards)):
  thread = Thread(target = ready, args = (shards[shard_idx]['ShardId'], ))
  thread.start()
```

Press Enter again and you should see this:

```
Thread-1 ready to process shard shardId-000000000000
Thread-2 ready to process shard shardId-000000000001
```

Your threads' numbers may be different. This is the basic pattern for the rest of our monitoring application: we will start a thread to process each shard, and each thread will run our monitoring code. One thing to note is that our application would have to be restarted if more shards were added (or indeed taken away), to create threads for the new shards. This is a limitation of working with Kinesis at such a low level. If we were using one of the Kinesis Client Libraries, changes in our stream's population of shards would be handled for us transparently.

Now we need a loop in each thread to handle reading events from each shard via shard iterators. The loop will be broadly the same as that of figure 4.11:

- Get an initial shard iterator for the given shard
- Use the shard iterator to read a batch of events from the shard
- Use the returned next shard iterator to read the next batch of events

We'll build this loop inside a Python class, imaginatively called `ShardReader`, which will run on its supplied thread. Type the following at the Python interpreter:

```
import time
from boto.kinesis.exceptions import ProvisionedThroughputExceededException

class ShardReader(Thread):
  def __init__(self, name, stream_name, shard_id):
    super(ShardReader, self).__init__(None, name)
    self.name = name
    self.stream_name = stream_name
    self.shard_id = shard_id
  def run(self):
    try:
      next_iterator = conn.get_shard_iterator(StreanName=self.stream_name,
        ShardId=self.shard_id,
      ShardIteratorType='TRIM_HORIZON')['ShardIterator']
      while True:
        response = conn.get_records(ShardIterator=next_iterator, Limit=10)
        for event in response['Records']:
          print(f"{self.name} read event {event['PartitionKey']}")
        next_iterator = response['NextShardIterator']
        time.sleep(5)
    except ProvisionedThroughputExceededException as ptee:
      print 'Caught: {}'.format(ptee.message)
      time.sleep(5)
```

We'll kick off the threads in a way similar to before:

```
for shard in shards:
  shard_id = shard['ShardId']
  reader_name = f'Reader-{shard_id}'
  reader = ShardReader(reader_name, 'events', shard_id)
  reader.start()
```

Press Enter and, assuming that your agent from section 4.1 is still running in another console, you should see something like the following:

```
Reader-shardId-000000000000 read event 481d142d-60f1-4d68-9bd6-d69eec5ff6c0
Reader-shardId-000000000001 read event 3486558d-163d-4e42-8d6f-c0fb91a9e7ec
Reader-shardId-000000000001 read event c3cd28b8-9ddc-4505-a1ce-193514c28b57
Reader-shardId-000000000000 read event f055a8bb-290c-4258-90f0-9ad3a817b26b
```

There will be a burst of events to start with. This is because both threads are catching up with the historic contents of each shard—all events dating from the so-called TRIM_HORIZON onward. After the backlog is cleared, the output should settle down to an event every 10 seconds or so, matching the rate at which our agent is producing readings. Note the Reader-shardId- prefixes on each of the output messages: these tell us which ShardReader each event is being read by. The allocation is random because each event is partitioned based on its event ID, which as a UUID is essentially random.

Let's see what happens if we stop the loop and then paste the same code in to start processing again:

```
Reader-shardId-000000000001 read event 3486558d-163d-4e42-8d6f-c0fb91a9e7ec
Reader-shardId-000000000000 read event 481d142d-60f1-4d68-9bd6-d69eec5ff6c0
Reader-shardId-000000000001 read event c3cd28b8-9ddc-4505-a1ce-193514c28b57
Reader-shardId-000000000000 read event f055a8bb-290c-4258-90f0-9ad3a817b26b
```

Compare the partition keys (remember, these are the event IDs) to the previous run, and you'll see that processing has restarted from the beginning of the shard. Our processing app is a goldfish; it has no memory of what events it has read on previous runs. Again, this is something that higher-level frameworks like the Kinesis Client Library handle: they allow you to "checkpoint" your progress against a stream by using persistent storage such as Amazon's DynamoDB.

Now that we have a stream processing framework in place, all that is left is to check the available disk space as reported in each event and detect if the available space drops below 20%. We need a function that takes a reading from our agent and generates an incident as required. Here is a function that does exactly that:

```
def detect_incident(event):
  decoded = json.loads(event)
  passed = None, None
  try:
    server = decoded['on']['server']['hostname']
    metrics = decoded['direct_object']['filesystem_metrics']
    pct_avail = metrics['available'] * 100 / metrics['size']
    return (server, pct_avail) if pct_avail <= 20 else passed
  except KeyError:
    return passed
```

Our function checks whether the proportion of available disk space is 20% or less, and if so, it returns the server's hostname and the proportion of available disk space in a tuple. If the check passes, we return a None, indicating that there is no action to take. We also tolerate any KeyError by returning a None, in case other event types are added to this stream in the future.

Let's test our new function in the Python interpreter with a valid event and an empty event:

```
detect_incident('{}')
(None, None)
detect_incident('{"on": {"server": {"hostname": "ulp"}},
 "direct_object": {"filesystem_metrics": {"available": 150, "free":
 100, "size": 1000}}, "verb": "read", "at": "2018-11-01T09:02:31.675773",
 "id": "b5ed136d-c879-4f2e-a9ce-a43313ce13c6", "subject": {"agent":
 {"version": "0.1.0"}}}')
(u'ulp', 15.0)
```

Good: the first call to detect_incident returned a (None, None) tuple, while the second call successfully detected that the server with hostname ulp has only 15% disk space available.

That's all the code that we need for our monitoring application. In the following listing, we consolidate everything into a single file, monitor.py.

Listing 4.2   monitor.py

```
#!/usr/bin/env python

import json, time, boto3
from threading import Thread
from boto.kinesis.exceptions import ProvisionedThroughputExceededException

class ShardReader(Thread):
  def __init__(self, name, stream_name, shard_id):
    super(ShardReader, self).__init__(None, name)
    self.name = name
    self.stream_name = stream_name
    self.shard_id = shard_id

  @staticmethod
  def detect_incident(event):
    decoded = json.loads(event)
    passed = None, None
    try:
      server = decoded['on']['server']['hostname']
      metrics = decoded['direct_object']['filesystem_metrics']
      pct_avail = metrics['available'] * 100 / metrics['size']
      return (server, pct_avail) if pct_avail <= 20 else passed
    except KeyError:
      return passed

  def run(self):
    try:
      next_iterator = conn.get_shard_iterator(StreamName=self.stream_name,
```

```
            ShardId=self.shard_id,
      ShardIteratorType='TRIM_HORIZON')['ShardIterator']
        while True:
          response = conn.get_records(ShardIterator=next_iterator, Limit=10)
          for event in response['Records']:
            print(f"{self.name} read event {event['PartitionKey']}")
            s, a = self.detect_incident(event['Data'])
            if a:
              print(f'{s} has only {a}% disk available!')
          next_iterator = response['NextShardIterator']
          time.sleep(5)
      except ProvisionedThroughputExceededException as ptee:
        print(f'Caught: {ptee.message}')
        time.sleep(5)

if __name__ == '__main__':
  session = boto3.Session(profile_name="ulp")
  conn = session.client("kinesis", region_name="eu-west-1")
  stream = conn.describe_stream(StreamName='events')
  shards = stream['StreamDescription']['Shards']

  threads = []
  for shard in shards:
    shard_id = shard['ShardId']
    reader_name = f'Reader-{shard_id}'
    reader = ShardReader(reader_name, 'events', shard_id)
    reader.start()
    threads.append(reader)

  for thread in threads:
    thread.join()
```

**Pass the event's JSON payload to detect_incident.**

**Print our incident if detected.**

**Create a list to hold all of our ShardReaders.**

**Wait for all ShardReaders to complete, i.e. loop indefinitely.**

Make the monitor.py file executable and run it:

```
chmod +x monitor.py
./monitor.py
Reader-shardId-000000000001 read event 3486558d-163d-4e42-8d6f-c0fb91a9e7ec
Reader-shardId-000000000000 read event 481d142d-60f1-4d68-9bd6-d69eec5ff6c0
Reader-shardId-000000000001 read event c3cd28b8-9ddc-4505-a1ce-193514c28b57
Reader-shardId-000000000000 read event f055a8bb-290c-4258-90f0-9ad3a817b26b
...
```

Our monitoring application is now running, reading each event in both shards of our events stream and reporting back on our server's disk usage. Unless you have a disk drive that's as cluttered as mine, chances are that the disk usage alert isn't firing. We can change that by temporarily creating an arbitrarily large file on our hard drive, using the fallocate command.

A quick filesystem check in my Vagrant virtual machine suggests that I have around 35 gigabytes available:

```
$ df -h
Filesystem      Size  Used Avail Use% Mounted on
/dev/sda1        40G    5G 34.8G  12% /
```

I created a temporary file sized at 30 gigabytes:

```
$ fallocate -l 30G /tmp/ulp.filler
$ df -h
Filesystem      Size  Used Avail Use% Mounted on
/dev/sda1        40G   35G  4.8G  88% /
```

And then I switch back to the terminal running monitor.py and see this:

```
Reader-shardId-000000000000 read event 097370ea-23bd-4225-ae39-fd227216e7d4
Reader-shardId-000000000001 read event e00ecc7b-1950-4e1a-98cf-49ac5c0f74b5
ulp has only 11% disk available!
Reader-shardId-000000000000 read event 8dcfe5ba-e14b-4e60-8547-393c20b2990a
ulp has only 11% disk available!
```

Great—the alert is firing! Every new filesystem metrics event being emitted by our agent.py is now triggering an alert in monitor.py. We can switch this off just as easily:

```
$ rm /tmp/ulp.filler
```

And switch back to our other terminal:

```
Reader-shardId-000000000001 read event 4afa8f27-3b62-4e23-b0a1-14af2ff1bfe1
ulp has only 11% disk available!
Reader-shardId-000000000000 read event 49b13b61-120d-44c5-8c53-ef5d91cb8795
Reader-shardId-000000000000 read event 8a3bf478-d211-49ab-8504-a0adae5a6a50
Reader-shardId-000000000000 read event 9d9a9b02-dea3-4ba1-adc9-464f4f2b0b31
```

So, this completes our systems monitoring example. To avoid incurring further AWS costs, you must now delete the events stream from the Kinesis screen in your AWS UI, as in figure 4.12.

To recap: we have written a simple systems monitoring agent in Python that emits a steady stream of filesystem metrics onto an Amazon Kinesis stream. We have then written another Python application, again using boto, the Amazon Python SDK, which monitors the agent's event stream in Kinesis, looking for hard drives filling up. Although our systems monitoring example is a rather narrow one, hopefully you can see how this could be extended to a more generalized monitoring approach.

Working with a two-shard Kinesis stream at a low level by using the AWS CLI and boto has given you a good handle on how Kinesis is designed. In the coming chapters, we will work with higher-level tools such as the KCL and Apache Spark Streaming, and your understanding of how Kinesis is designed at this lower level will stand us in good stead for this.

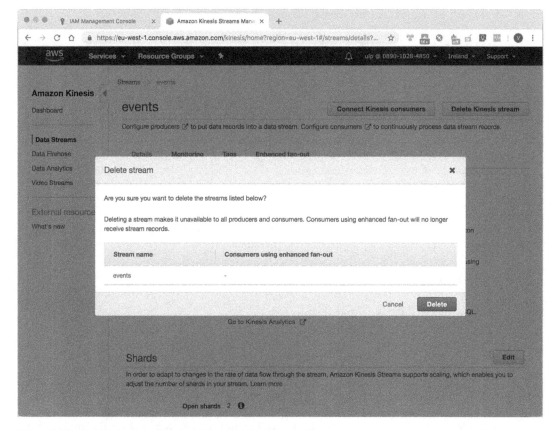

**Figure 4.12   Deleting our events stream by using the Kinesis UI**

## Summary

- Amazon Kinesis is a fully managed unified log service, available as part of the Amazon Web Services offering.
- Amazon Kinesis and Apache Kafka have differences in terminology, but the semantics of both technologies are extremely similar.
- A wide array of stream processing frameworks already support Amazon Kinesis, including Apache Spark Streaming, Apache Storm, and Amazon's own Kinesis Client Libraries (KCLs) for Java and Python.
- We can set up an identity and access management (IAM) user in AWS, and assign a managed policy to that user to give full permissions on Kinesis.
- Using the AWS CLI, we can start writing events to a Kinesis stream, and read those same events out by using an abstraction called a shard iterator.
- Systems monitoring is a new use case for unified log processing. We can model systems monitoring events by using our standard grammatical approach, and create agents that run on servers and push monitoring events into our unified log.

- We can build a simple systems monitoring agent in Python by using the Python SDK for AWS, called boto. This agent runs in an infinite loop, regularly emitting events that consist of readings of filesystem statistics from the local server.
- We can build a low-level monitoring application in Python by using Python threads and boto. Each thread reads all events from a single Kinesis shard by using shard iterators, monitors the reported disk usage, and raises an incident if disk usage reaches 80%.
- Higher-level tools such as the KCL will handle the basics of Kinesis stream processing for us, including distribution of work across multiple servers and checkpointing in a database our progress against each shard.

# Stateful stream processing

In chapter 3, we introduced the idea of processing continuous event streams and implemented a simple application that processed individual shopping events from the Nile website. The app we wrote did a few neat things: it read individual events from Kafka, filtered out bad input events, enriched the event with location information, and finally wrote the newly filtered and enriched event back out to Kafka.

Chapter 3's app was relatively simple because it operated on only a single event at a time: it read each individual event off a Kafka topic, and then decided whether it would either filter the event (discard it), or enrich the event and write that enriched event back to a new Kafka topic. In the terminology introduced in chapter 3, our app was performing *single-event processing*, whereby one input event generates zero or more output events, in contrast to what we call *multiple-event processing*, whereby one or more input events generates zero or more output events.

This chapter is all about multiple-event processing—or as we will start calling it (for reasons I'll explain soon)—*stateful stream processing*. In this chapter, we will write an application that generates outgoing events based on multiple incoming events. Continuing our employment at fictitious online retailer Nile, this time we will be striving to improve the online shopping experience, by detecting whenever a Nile shopper has abandoned their shopping cart.

As we hinted in chapter 3, processing multiple events at a time is more complex than processing single events. We need to maintain some form of state to keep track of shopper behavior across multiple events; this brings with it attendant challenges, such as distributing the processing and the state safely over multiple servers. To meet these challenges, we will introduce *stream processing frameworks* at a high level.

We will implement our abandoned shopping cart detector in Java by using the Apache Samza stream processing framework. Samza is not the most featured or well-known stream processing framework; the API it provides is relatively basic, slightly reminiscent of Hadoop's original MapReduce API in Java. But Samza's simplicity is also a strength: it will make it easier for us to see and understand the essential stateful nature of this stream processing job. Writing this job with Samza should give you the confidence to try out the newer and "buzzier' frameworks like Spark Streaming and Flink.

But first we will introduce our new business challenge and outline its stream processing requirements.

## 5.1 Detecting abandoned shopping carts

Remember: we currently work for a sells-everything e-commerce website, called Nile. The management team at Nile wants the company to become much more dynamic and responsive by reacting to customer behavior in a timely and effective fashion. This will be a great opportunity for us to implement *stateful stream processing*.

### 5.1.1 What management wants

As part of their goal of creating a more dynamic and responsive business, the management team at Nile has identified a key opportunity around identifying and reacting to shopper-abandoned shopping carts. A shopping cart is defined as *abandoned* when a shopper adds products to their shopping cart but doesn't continue through to checkout.

For online retailers like Nile, it is worthwhile contacting shoppers who have abandoned their carts and asking if they would like to complete their orders. Timing is everything here: it's important to identify and react to abandoned shopping carts quickly, but not so quickly that a shopper feels pestered or rushed; UK handbag company Radley released a study showing that 30 minutes after abandonment is the optimal time to get in touch.[1]

---

[1] Details of this study are available at http://d34w0339mx0ifp.cloudfront.net/global/images/uploads/2013/11/Radley-Client-Story.pdf.

Typically, an online retailer responds to an abandoned shopping cart by emailing the shopper, or by showing the shopper retargeting ads on other websites, but we don't need to worry about the exact response mechanism. The important thing is to define a new event, *Shopper abandons cart*, and generate one of these new events whenever we detect an abandoned cart. Nile's data engineers can then read these new events from the relevant stream and decide how to handle them.

### 5.1.2    Defining our algorithm

How do we detect an abandoned shopping cart? For the purposes of this chapter, let's use a simple algorithm:

- Our shopper adds a product to the cart.
- Derive a *Shopper abandons cart* event if 30 minutes pass without one of the following occurring:
  - *Our shopper adding any further products to cart*
  - *Our shopper placing an order*
- If our shopper adds a further product to the cart during the 30 minutes, restart the timer.
- If our shopper places an order during the 30 minutes, clear the timer.

Figure 5.1 presents two examples of applying this algorithm. This algorithm is not a particularly sophisticated one, but it should help Nile get started tackling its abandoned shopping carts problem, and it can always be refined further later.

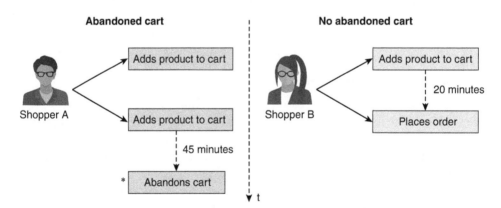

**Figure 5.1    On the left side, Shopper A has added two products to the shopping cart, and then 45 minutes have passed without any further activity, so we can derive a *Shopper abandons cart* event. On the right side, Shopper B has added a product to the shopping cart and checked out within 20 minutes, not abandoning the cart.**

### 5.1.3  *Introducing our derived events stream*

When an abandoned cart is detected, a new *Shopper abandons cart* event is generated, per the preceding algorithm. But where should we write this new event to? We have two options:

- We write it back to the `raw-events-ch05` topic, colocating it with the original shopper events that were fed into the algorithm.
- We write it to a new Kafka topic, called `derived-events-ch05`.

Both approaches have their merits. Writing the new event back to `raw-events-ch05` is simple to reason about: all events can now be found in a single stream, regardless of the process that generated them. On the other hand, writing to a separate stream, `derived-events-ch05`, makes it clear that these new events are *second-order events*, derived from the original events. A data engineer who cares only about this event (for example, to send abandoned shopping cart emails) can read this new stream and ignore the original stream.

For this chapter, we will opt for the second approach and write our *Shopper abandons cart* events to a new Kafka topic, called `derived-events-ch05`. Figure 5.2 shows our new stream processing job, reading events from the `raw-events-ch05` topic, and writing new events to a `derived-events-ch05` topic.

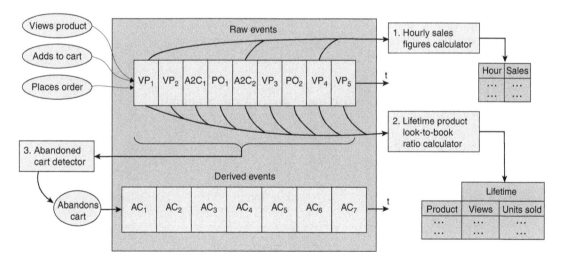

**Figure 5.2**  **Our abandoned shopping cart detector consumes events from the `raw-events-ch05` topic in Kafka and generates new *Shopper abandons cart* events to write back to a new Kafka topic, called `derived-events-ch05`.**

There is a lot to take in here: we have defined an algorithm to meet Nile's goals around abandoned shopping carts and introduced a second Kafka topic to receive the new events. Before we go any further, we need to model the various event types that we will be dealing with here.

## 5.2   *Modeling our new events*

Remember back in chapter 2, when we introduced our first e-commerce-related event for Nile, *Shopper views product?* Drawing on our standard definition of an event as *subject-verb-object*, we can identify three further discrete events required for tracking abandoned carts:

- *Shopper adds item to cart*—The shopper adds one of those products to the shopping basket. A product is added to the basket with a quantity of one or more attached.
- *Shopper places order*—The shopper checks out, paying for the items in the shopping basket.
- *Shopper abandons cart*—The derived event itself, representing the act of cart abandonment by the shopper.

Before we jump into writing our stream processing application, let's first *model* these three new event types. Although it's tempting to skip this step and dive into the coding, having clear definitions of the events that we will be working with will save us a lot of time and should prevent us from getting stuck down any coding cul de sacs.

### 5.2.1   *Shopper adds item to cart*

The *Shopper adds item to cart* event involves a Nile shopper adding a product to their shopping cart (also known as a *shopping basket*), specifying a quantity of that product as they do this. Let's break out the various components here:

- *Subject:* Shopper
- *Verb:* Adds
- *Direct object:* Item (consisting of product and quantity)
- *Indirect object:* Cart
- *Context:* Timestamp of this event

Figure 5.3 illustrates these components.

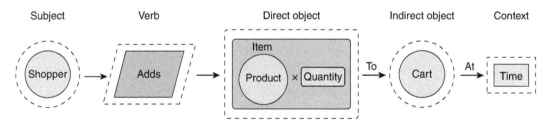

**Figure 5.3   Our shopper (again, subject) adds (verb) an item, consisting of a product and its quantity (direct object) to the shopping cart (indirect, aka prepositional, object) at a given time (context).**

### 5.2.2 Shopper places order

This event sounds simple, and it is:

- *Subject:* Shopper
- *Verb:* Places
- *Direct object:* Order
- *Context:* Timestamp of this event

The slight complexity is in modeling the order. This is a complicated entity: it needs to contain an order ID and a total order value, plus a list of items that were purchased in the order; each item should be a product and the quantity of that product ordered.

Putting it all together, we get the event drawn in figure 5.4.

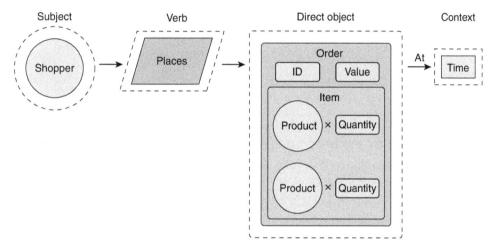

**Figure 5.4  Our shopper (always the subject) places (verb) an order (direct object) at a given time (context). The order contains ID and value attributes, plus an array of order items, each consisting of a product and quantity of that product.**

### 5.2.3 Shopper abandons cart

Now we come to our derived event. By *derived*, we mean an event that we are generating ourselves in our stream processing application, as opposed to an incoming raw event that we are simply consuming.

This event looks like this:

- *Subject:* Shopper
- *Verb:* Abandons
- *Direct object:* Cart (consisting of multiple items, each of a product and quantity)
- *Context:* Timestamp of this event

Our fourth and final event for this chapter is illustrated in figure 5.5.

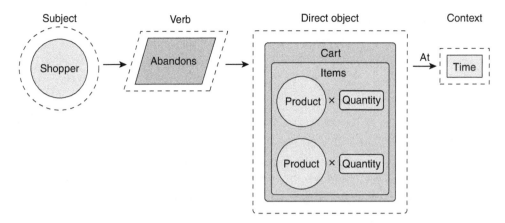

**Figure 5.5   Our shopper (subject) abandons (verb) their shopping cart (direct object) at a given time (context). The shopping cart contains an array of order items, each consisting of a product and quantity of that product.**

This completes the set of Nile e-commerce events required for our abandoned shopping cart detector. In the next section, we will look at our final building block: stateful stream processing.

## 5.3   *Stateful stream processing*

To detect abandoned shopping carts for Nile, we have to do some *stream processing* on the incoming events. Clearly, it's a little more complex than chapter 3, where we were able to work on only a single event at a time: our algorithm expects us to understand the flow of multiple events in a sequence over time. The key building block for processing multiple events like this is state.

### 5.3.1   *Introducing state management*

If single-event processing is a little like being a goldfish, multiple-event processing is like being an elephant. The goldfish stream processor can forget each event as soon as it has processed it, whereas the elephant stream processor requires a memory of prior events in order to generate its output. We can call this memory of prior events *state* and say that multiple-event processing is therefore *stateful*. Figure 5.6 illustrates this slightly tortuous metaphor.

What kind of technology should we be using to implement this state in our stream processing application? There are lots of options, but they largely boil down to three:

- *In-process memory*—A mutable variable or variables available inside the stream processing application's own code
- *Local data store*—Some kind of simple data store (for example, LevelDB, RocksDB, or SQLite) that is local to the server or container running this instance of the stream processing application
- *Remote data store*—Some kind of database server or cluster (for example, Cassandra, DynamoDB, or Apache HBase) hosted remotely

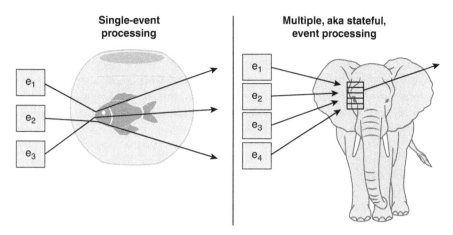

**Figure 5.6  Our goldfish-like single-event processor can process each event and then immediately forget about it. By contrast, our elephant-like stateful event processor must remember multiple events to perform aggregations, sorting, pattern detection, and suchlike.**

Figure 5.7 depicts these three options. Regardless of the specific approach, the important thing to understand is that processing multiple events needs state.

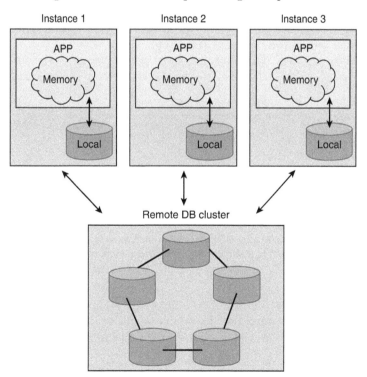

**Figure 5.7  A distributed stream-processing application (here with three instances) can keep state in in-process memory, in an instance-local data store, or in a remote data store—or even in a combination of all three.**

### 5.3.2   *Stream windowing*

We know that we need some form of state to process multiple events, but what exactly is going to be kept in that state? When we think of data processing, we are almost always thinking of processing a *bounded set* of data—a collection of data points that are of a somehow finite size. After all, even an elephant has a limit on how much they can remember. Here are some examples of data queries and transformations from a bounded set, with the bounded set in italics:

- Who is the fastest marathon runner *since records began?*
- How many T-shirts did we give away *in January 2018?*
- What are our *year-to-date* revenues?

Unfortunately for our purposes, a continuous event stream is *unterminated,* meaning that it has no end (and possibly no beginning either), so the first challenge in multiple-event processing is to come up with a sensible way of bounding a continuous event stream. We typically solve this problem by applying a *processing window* to our event stream.

At its simplest, processing on a continuous event stream can be put into discrete windows by using a timer, or heartbeat, function that runs at a regular interval. This timer function can contain custom code to apply whatever window or windows makes sense for the use case. Figure 5.8 illustrates this idea of slicing an unending event stream into specific windows for further processing.

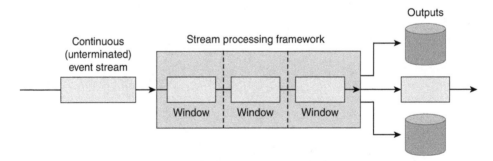

**Figure 5.8   A stream processing framework typically applies windowing to a continuous (unterminated) event stream to process it and generate meaningful outputs.**

In the case of our abandoned shopping cart, the processing window is well-defined: we are looking for shopping carts that have been abandoned by their owner for at least 30 minutes. We don't want to leave an abandoned shopping cart much longer than the required 30 minutes, so a timer checking each cart every 30 seconds would be ideal.

If stream windowing doesn't make a lot of sense yet, don't worry; we will be exploring it in more detail when we start building our abandoned cart detector. The key takeaway for now is that stream windowing is an important building block for stateful stream processing.

### 5.3.3 *Stream processing frameworks and their capabilities*

All this talk of processing multiple events, maintaining state, and creating stream windows probably sounds like a lot of work! To solve these and other problems in a repeatable, reliable way, a handful of *stream processing frameworks* have emerged. These frameworks exist to abstract away the mechanics of running stateful stream processing at scale, leaving developers to focus on the actual business logic that their jobs require.

Stream processing frameworks exhibit some or all of the following capabilities:

- *State management*—State is a key building block for processing multiple events at a time. Stream processing frameworks give you the option of storing state in some or all of the following: in-memory, in a local filesystem, or in a dedicated key-value store such as RocksDB or Redis.
- *Stream windowing*—As described previously, a stream processing framework provides one or more ways of expressing a *bounded window* for the event processing. This is typically time-based, although sometimes it can be based on a count of events instead.
- *Delivery guarantees*—All stream processing frameworks pessimistically track their progress, to ensure that every event is processed *at least once*. Some of these systems go further, adding in transactional guarantees to ensure that each event is processed only *exactly once*.
- *Task distribution*—A high-volume event stream consists of multiple Kafka topics or Amazon Kinesis streams, and requires multiple instances of the job, or tasks, to process it. Most stream processing frameworks are designed to be run on a sophisticated third-party scheduler such as Apache Mesos or Apache Hadoop YARN; some stream processing frameworks are also *embeddable*, meaning that they can be added as a library to a regular application (requiring bespoke scheduling).
- *Fault tolerance*—Failures happen regularly in the kind of large-scale distributed systems required to process high-volume event streams, and most frameworks have built-in mechanisms to automatically recover from these failures. Fault tolerance typically involves a distributed backup of either the incoming events or the generated state.

Let's see how some of the most popular stream processing frameworks relate to these capabilities.

### 5.3.4 *Stream processing frameworks*

A variety of stream processing frameworks have emerged over the past few years, with most of the popular ones having been successfully incubated by the Apache Software

Foundation. Table 5.1 introduces five of the most widely used frameworks and lays out their design relative to the capabilities introduced in the preceding section.

Table 5.1   A nonexhaustive list of stream processing frameworks

| Capability | Storm | Samza | Spark Streaming | Kafka Streams | Flink |
|---|---|---|---|---|---|
| State management | In-memory, Redis | In-memory, RocksDB | In-memory, filesystem | In-memory, RocksDB | In-memory, filesystem, RocksDB |
| Stream windowing | Time-, count-based | Time-based | Time-based (microbatch) | Time-based | Time-, count-based |
| Delivery guarantees | At least once, exactly once (Trident) | At least once | At least once, exactly once | At least once, exactly once | Exactly once |
| Task distribution | YARN or Mesos | YARN or embeddable | YARN or Mesos | Embeddable | YARN or Mesos |
| Fault tolerance | Record acks | Local, distributed snapshots | Checkpoints | Local, distributed snapshots | Distributed snapshots |

Let's go through each of these stream processing frameworks briefly.

### APACHE STORM

*Apache Storm* was the first of the "new wave" of stream processing frameworks. Storm was written by Nathan Marz at BackType; Twitter acquired BackType and open sourced Storm in 2011. As a pioneering piece of technology, Storm did things slightly differently than its successors:

- Fault tolerance is achieved by record acknowledgments with upstream record backups, rather than state snapshots or checkpoints.
- Rather than using a third-party scheduler, Storm originally created its own (Nimbus)—although Storm supports both YARN and Mesos now.
- Storm introduced a separate library, Storm Trident, to support exactly-once processing.

Storm was popularized by the book *Big Data* by Nathan Marz and James Warren (Manning) and continues to be widely used; Twitter's successor system, Heron, and Apache Flink (covered later in this section) both offer Storm-compatible APIs to ease adoption.

### APACHE SAMZA

*Apache Samza* is a stateful stream-processing framework from the team at LinkedIn who also originated Apache Kafka. As such, Samza has tight integration with Apache Kafka, using Kafka for the backing up of the state held in RocksDB databases. Samza has a relatively low-level API, letting you interact directly with Samza's stream windowing and state management features. Samza promotes building any required complex *stream-processing topology* out of many simpler Samza jobs, reading from and writing back to separate Kafka topics.

Samza jobs were originally designed to be run from the YARN application scheduler; more recently, Samza can also be embedded as a library inside a regular application, letting you choose your own scheduler (or no scheduler). Samza has also outgrown its original inter-dependence with Kafka, now also supporting Amazon Kinesis.

Samza continues to evolve, is a great teaching tool for stream processing, and is used by major companies. But Kafka Streams (covered later in this section) has perhaps stolen some of Samza's thunder.

### APACHE SPARK STREAMING

*Spark Streaming* is an extension of the Apache Spark project for working with continuous event streams. Technically speaking, Spark Streaming is a *microbatch* processing framework, not a *stream* processing framework: Spark Streaming slices the incoming event stream into microbatches, which it then feeds to the standard Spark engine for processing.

Spark Streaming's microbatching gives us exactly-once processing "for free" and lets us reuse our existing Spark experience and code, should we have it. But it comes at a cost: the microbatching increases latency and reduces our flexibility around stream windowing and scaling.

Spark Streaming is not tied to any one stream technology: almost any incoming stream technology can be sliced and fed into Spark in microbatches; out-of-the-box Spark Streaming supports Flume, Kafka, Amazon Kinesis, files, and sockets, and you can write your own *custom receiver* if you like.

For readers interested in digging further into Spark Streaming, *Spark in Action* by Petar Zečević and Marko Bonaći, and *Streaming Data* by Andrew G. Psaltis, both published by Manning, are great resources.

### APACHE KAFKA STREAMS

*Kafka Streams* is a stream processing library for Apache Kafka, conceived by the LinkedIn Kafka team after they decamped LinkedIn for Kafka startup Confluent. Kafka Streams has shared conceptual DNA with Samza, but is on a somewhat different trajectory:

- Kafka Streams was designed from the start as a library to be embedded in your own application code; it doesn't support any existing scheduler such as YARN or Mesos.
- Kafka Streams is closely tied to Kafka; Kafka Streams uses Kafka for all aspects of its state management, fault tolerance, and delivery guarantees; you cannot use Kafka Streams with any other stream technology.
- Although Kafka Streams does expose a low-level API, it also has a higher-level query API that will be more familiar to Spark users.

Although a young project, Kafka Streams is quickly gaining serious developer mindshare as a first-class citizen within the Kafka ecosystem. Confluent is positioning Kafka

Streams as a toolkit for asynchronous or event-driven microservices, as distinct from the more analytics or data science use cases that Spark is noted for.

Another great Manning book to dig into Kafka Streams is *Kafka Streams in Action* by William P. Bejeck Jr.

**APACHE FLINK**

*Apache Flink* is a relative newcomer but rapidly emerging as a credible challenger to Spark and Spark Streaming. Unlike Spark, which takes a batch model and handles streaming use cases via microbatching, Flink was built from the start as a streaming engine: it handles batch data by treating it as a bounded stream, one with a beginning and an end.

Like Spark, Flink is not tied to Kafka. Flink supports various other data sources and sinks, including Amazon Kinesis. Flink has sophisticated stream windowing capabilities, and supports exactly-once processing and a rich query API; Flink is also closely following the Apache Beam project, which aims to provide a standard, full-featured API for interacting with stream processing frameworks.

This concludes our introduction to five major stream-processing frameworks, but how do we choose between these for our work at Nile?

### 5.3.5   *Choosing a stream processing framework for Nile*

As mentioned at the beginning of this chapter, we are going to use Apache Samza to build our abandoned cart detector. Samza is a full-blown stream processing framework, but unlike the other tools we've introduced, Samza makes no effort to abstract or otherwise hide away the stateful nature of the event processing involved: with Samza, you will be working directly with a simple key-value data store to keep track of events that you have already processed.

What is a *key-value store*? It's a super-simple database in which you store a single *value* (an array of bytes) against an individual unique *key* (also represented by an array of bytes); the key and the value can be any value that you like. A key-value store is a crude tool, but an effective one. It will give us the state we need to track shopper behavior across multiple events. Samza uses RocksDB, an embeddable persistent (versus in-memory only) key-value store, created by Facebook.

Samza has other tricks up its sleeves. The core of a Samza job consists of just two functions:

- process()—Called for each incoming event
- window()—Called on a regular, configurable interval

Both functions have access to the key-value store, plus any mutable state defined in the job; this lets them effectively communicate with each other.

If other stream processing frameworks are fishing sticks, Samza is a fishing rod and tackle: the way that Apache Samza exposes the underlying key-value store as a first-class entity for direct manipulation makes Samza an excellent teaching tool for stateful stream processing. Similarly, the no-nonsense `process()` and `window()` functions make it easy to visualize exactly what is going on in our stream processing job at any moment.

## 5.4    Detecting abandoned carts

A recap: we are writing a stateful stream-processing job that will search the incoming stream, looking for a specific pattern of events, and when that pattern is found, the job will emit a new event, *Shopper abandons cart*. We are going to implement this job in Apache Samza. Let's get started.

### 5.4.1    Designing our Samza job

To detect abandoned carts, we need a way of encapsulating shopper behavior in Samza's key-value store. We must represent each observed shopper's current *state* in the key-value store and keep this state up-to-date so that we can generate a *Shopper abandons cart* event as soon as that behavior is detected.

With a key-value store, it is completely up to us how we design our *keyspace*—the meaning and layout of keys (and thus their values) in the database. In this case, we can use a pair of namespaced keys to maintain our shopper's current state in Samza, like so:

- `<shopper>-ts` should be kept up-to-date with the timestamp at which our job saw the most recent *Shopper adds item to cart* event for the given shopper.
- `<shopper>-cart` should be kept up-to-date with the current contents of the shopper's cart, based on aggregating all *Shopper adds item to basket* events.

It's important that we don't pester Nile's customers about abandoned carts. So, we should delete both keys from Samza (effectively "resetting" tracking for this user) as soon as we see a *Shopper places order* event; this should be done regardless of which products were purchased in that order. We should also delete these keys immediately after sending a new *Shopper abandons cart* event, so we don't send the event twice.

Putting this together, we have two events that we are interested in for our `process()` function:

- *Shopper adds item to cart*—For tracking the shopper's cart state and the time when they were last active with their cart
- *Shopper places order*—For telling us to "reset" tracking for this user

For our `window()` function, we will scan through the whole key-value store, looking for shoppers who were last active more than 30 minutes ago, based on the `<shopper>-ts` values. Whenever we find one, we will generate a new *Shopper abandons cart* event, containing the cart contents for this shopper as fetched from `<shopper>-cart`.

There's a lot to take in here. Don't worry—it should become clearer after you see the Java task code. In the meantime, figure 5.9 sets out the overall design of this job.

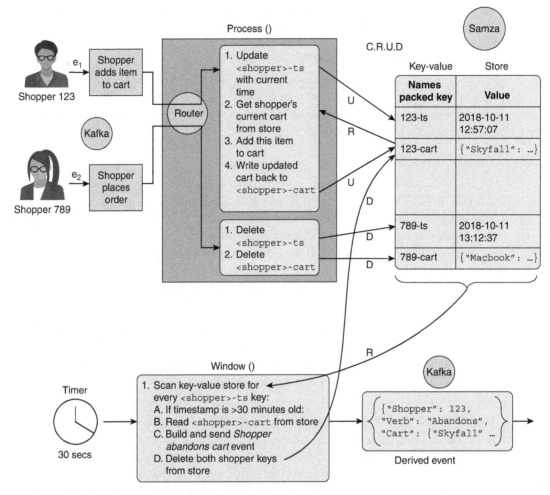

**Figure 5.9   Our `process()` function parses incoming events from Nile's shoppers and updates the Samza key-value store to track the shoppers' behavior. The `window()` function then runs regularly to scan the key-value store and identify shoppers who have abandoned their carts. A *Shopper abandons cart* event is then emitted for those shoppers.**

### 5.4.2   *Preparing our project*

Unfortunately, a lot of ceremony exists around setting up a new Samza project, which is a distraction from the actual stream processing we want to implement. Happily, we can skip most of this ceremony by starting from the Samza's team's own Hello World project. This will serve as our cuckoo's nest. Let's check out the project like so:

```
$ git clone https://git.apache.org/samza-hello-samza.git nile-carts
$ cd nile-carts
$ git checkout f488927
Note: checking out 'f488927'.`
    ...
```

Checking out the specific commit ensures that we are using the release of Hello World that works with Samza version 0.14.0.

Let's clean up the folder structure a little:

```
$ rm wikipedia-raw.json
$ rm src/main/config/*.properties
$ rm -rf src/main/java/samza
```

Apache projects use a Maven plugin called Rat. Using this with our project will cause problems, so let's delete this from the Maven pom.xml file:

```
$ sed -i '' '257,269d' pom.xml
```

Finally, let's rename the artifact to prevent confusion later. Edit the pom.xml file in the root and change the `artifactId` element on line 29 from `hello-samza` to the following value:

```
<artifactId>nile-carts</artifactId>
```

Similarly, edit the file src/main/assembly/src.xml and replace the `include` element on line 69 from `hello-samza` to the following value:

```
<include>org.apache.samza:nile-carts</include>
```

That's enough feathering of the nest. Let's move on to configuring our job.

### 5.4.3 Configuring our job

Although Samza jobs are written in Java, they are typically assembled with a Java properties file that tells Samza exactly how to run the file. Let's create this configuration file now. Open your editor and create a file at this path:

```
src/main/config/nile-carts.properties
```

Populate your new properties file with the following configuration.

**Listing 5.1  nile-carts.properties**

```
# Job
job.factory.class=org.apache.samza.job.yarn.YarnJobFactory
job.name=nile-carts
job.coordinator.system=kafka
job.coordinator.replication.factor=1

# YARN
yarn.package.path=file://${basedir}/target/${project.artifactId}-
➥ ${pom.version}-dist.tar.gz

# Task
task.class=nile.tasks.AbandonedCartStreamTask
task.inputs=kafka.raw-events-ch05
task.window.ms=30000
```

```
# Serializers
serializers.registry.json.class=org.apache.samza.serializers.
JsonSerdeFactory
serializers.registry.string.class=org.apache.samza.serializers.
⟼ StringSerdeFactory

# Systems
systems.kafka.samza.factory=org.apache.samza.system.kafka.KafkaSystemFactory
systems.kafka.samza.msg.serde=json
systems.kafka.consumer.zookeeper.connect=localhost:2181/
systems.kafka.producer.bootstrap.servers=localhost:9092
systems.kafka.consumer.auto.offset.reset=largest
systems.kafka.producer.metadata.broker.list=localhost:9092
systems.kafka.producer.producer.type=sync
systems.kafka.producer.batch.num.messages=1

# Key-value storage
stores.nile-carts.factory= org.apache.samza.storage.kv
     .RocksDbKeyValueStorageEngineFactory
stores.nile-carts.changelog=kafka.nile-carts-changelog
stores.nile-carts.changelog.replication.factor=1
stores.nile-carts.key.serde=string
stores.nile-carts.msg.serde=string
stores.nile-carts.write.batch.size=0
stores.nile-carts.object.cache.size=0
```

To go through each of the configuration blocks in turn:

- The Job section tells us that this is the nile-carts job, and will be run on YARN.
- The YARN section defines the location of our Samza job package for YARN.
- The Task section specifies the Java class that will contain the job's stream processing logic, which we'll write in the next section. The Java class will process events coming from the Kafka topic raw-events-ch05, and it will have a processing window of 30 seconds (30,000 milliseconds).
- The Serializers section declares two *serdes* (serializers-deserializers) that will let us read and write strings and JSON in our job.
- The Systems section configures Kafka for our job. We specify that events we consume and produce from Samza will be in JSON format.
- The Key-value storage section configures our local state. Our keys and values in the key-value store will all be strings.

And that's it for the configuration—now on to the code.

### 5.4.4  *Writing our job's Java task*

Remember that this Samza job needs to generate a stream of well-structured *Shopper abandons cart* events. Rather than jumping straight into our task code, let's first create a base Event class, and then extend it for our AbandonedCartEvent.

In your editor, add the code in the following listing into a file at this path:

```
src/main/java/nile/events/Event.java
```

**Listing 5.2   Event.java**

```
package nile.events;
                                        Using the
import org.joda.time.DateTime;          JodaTime library
import org.joda.time.DateTimeZone;
import org.joda.time.format.DateTimeFormat;
import org.joda.time.format.DateTimeFormatter;
import org.codehaus.jackson.map.ObjectMapper;

public abstract class Event {           Our abstract Event
                                        class. Contains subject,
  public Subject subject;               verb, and context fields.
  public String verb;
  public Context context;

  protected static final ObjectMapper MAPPER = new ObjectMapper();
  protected static final DateTimeFormatter EVENT_DTF = DateTimeFormat
    .forPattern("yyyy-MM-dd'T'HH:mm:ss").withZone(DateTimeZone.UTC);

  public Event(String shopper, String verb) {
    this.subject = new Subject(shopper);
    this.verb = verb;
    this.context = new Context();
  }

  public static class Subject {
    public final String shopper;         Our subject is a shopper,
                                         identified by a cookie
    public Subject() {                    value in a string.
      this.shopper = null;
    }

    public Subject(String shopper) {
      this.shopper = shopper;
    }
  }

  public static class Context {
    public final String timestamp;
                                          Our context's timestamp
    public Context() {                    will be set when the
      this.timestamp = EVENT_DTF.print(   event is constructed.
        new DateTime(DateTimeZone.UTC));
    }
  }
}
```

Our abstract Event class is just a helper for modeling an event that conforms to the structure of *subject-verb-object*. We can extend it to create the AbandonedCartEvent class that our job will use to send well-formed *Shopper abandons cart* events. Stub a file at this path:

```
src/main/java/nile/events/AbandonedCartEvent.java
```

Add in the code in the following listing.

**Listing 5.3    AbandonedCartEvent.java**

```
package nile.events;

import java.io.IOException;
import java.util.List;
import java.util.ArrayList;
import java.util.Map;
import org.joda.time.DateTime;
import org.joda.time.DateTimeZone;
import org.codehaus.jackson.type.TypeReference;
import nile.events.Event;

public class AbandonedCartEvent extends Event {
  public final DirectObject directObject;

  public AbandonedCartEvent(String shopper, String cart) {
    super(shopper, "abandon");
    this.directObject = new DirectObject(cart);
  }

  public static final class DirectObject {
    public final Cart cart;

    public DirectObject(String cart) {
      this.cart = new Cart(cart);
    }

    public static final class Cart {

      private static final int ABANDONED_AFTER_SECS = 1800;

      public List<Map<String, Object>> items =
        new ArrayList<Map<String, Object>>();

      public Cart(String json) {
        if (json != null) {
          try {
            this.items = MAPPER.readValue(json,
              new TypeReference<List<Map<String, Object>>>() {});
          } catch (IOException ioe) {
            throw new RuntimeException("Problem parsing JSON cart", ioe);
          }
        }
      }

      public void addItem(Map<String, Object> item) {
        this.items.add(item);
      }

      public String asJson() {
        try {
          return MAPPER.writeValueAsString(this.items);
```

> We will classify a cart as abandoned if no event activity occurs for 30 minutes.

> Adds an item to the cart. Note that we don't bother "de-duplicating" multiple additions of the same product.

> Converts the items in our cart to JSON, ready for writing to the key-value store

```
      } catch (IOException ioe) {
        throw new RuntimeException("Problem writing JSON cart", ioe);
      }
    }

    public static boolean isAbandoned(String timestamp) {
      DateTime ts = EVENT_DTF.parseDateTime(timestamp);
      DateTime cutoff = new DateTime(DateTimeZone.UTC)
        .minusSeconds(ABANDONED_AFTER_SECS);
      return ts.isBefore(cutoff);
    }
  }
 }
}
```

**Checks whether the supplied timestamp counts as abandoned (if more than 30 minutes old)**

Our `AbandonedCartEvent` gives us an easy way of generating a new *Shopper abandons cart* event, ready for our Samza job to emit whenever an abandoned cart is detected. The most interesting aspect is the `Cart` inner class, which contains a helper method used to add items to a cart whenever a *Shopper adds item to cart* event occurs. Meanwhile, the `isAbandoned()` helper function will tell us whether a given shopping cart has been abandoned.

With our *Shopper abandons cart* event defined, we are now ready to write the Java `StreamTask` that makes up the core of our Samza job. Back in your editor, create a file at this path:

`src/main/java/nile/tasks/AbandonedCartStreamTask.java`

And for the last time in this chapter, add in the code in the following listing.

**Listing 5.4  AbandonedCartsStreamTask.java**

```
package nile.tasks;

import java.util.HashMap;
import java.util.HashSet;
import java.util.Map;
import java.util.Set;
import org.apache.samza.config.Config;
import org.apache.samza.storage.kv.KeyValueStore;
import org.apache.samza.storage.kv.KeyValueIterator;
import org.apache.samza.storage.kv.Entry;
import org.apache.samza.system.IncomingMessageEnvelope;
import org.apache.samza.system.OutgoingMessageEnvelope;
import org.apache.samza.system.SystemStream;
import org.apache.samza.task.InitableTask;
import org.apache.samza.task.MessageCollector;
import org.apache.samza.task.StreamTask;
import org.apache.samza.task.TaskContext;
import org.apache.samza.task.TaskCoordinator;
import org.apache.samza.task.WindowableTask;
import nile.events.AbandonedCartEvent;
import nile.events.AbandonedCartEvent.DirectObject.Cart;
```

```
public class AbandonedCartStreamTask
  implements StreamTask, InitableTask, WindowableTask {

  private KeyValueStore<String, String> store;

  public void init(Config config, TaskContext context) {
    this.store = (KeyValueStore<String, String>)
      context.getStore("nile-carts");
  }

  @SuppressWarnings("unchecked")
  @Override
  public void process(IncomingMessageEnvelope envelope,
    MessageCollector collector, TaskCoordinator coordinator) {

    Map<String, Object> event =
      (Map<String, Object>) envelope.getMessage();
    String verb = (String) event.get("verb");
    String shopper = (String) ((Map<String, Object>)
      event.get("subject")).get("shopper");

    if (verb.equals("add")) {
      String timestamp = (String) ((Map<String, Object>)
        event.get("context")).get("timestamp");

      Map<String, Object> item = (Map<String, Object>)
        ((Map<String, Object>) event.get("directObject")).get("item");
      Cart cart = new Cart(store.get(asCartKey(shopper)));
      cart.addItem(item);

      store.put(asTimestampKey(shopper), timestamp);
      store.put(asCartKey(shopper), cart.asJson());

    } else if (verb.equals("place")) {
      resetShopper(shopper);
    }
  }

  @Override
  public void window(MessageCollector collector,
    TaskCoordinator coordinator) {

    KeyValueIterator<String, String> entries = store.all();
    while (entries.hasNext()) {
      Entry<String, String> entry = entries.next();
      String key = entry.getKey();
      String value = entry.getValue();
      if (isTimestampKey(key) && Cart.isAbandoned(value)) {
        String shopper = extractShopper(key);
        String cart = store.get(asCartKey(shopper));

        AbandonedCartEvent event =
          new AbandonedCartEvent(shopper, cart);
        collector.send(new OutgoingMessageEnvelope(
          new SystemStream("kafka", "derived-events-ch05"), event));
```

For an add-to-cart event, fetch the current cart for this shopper from the key-value store, add this new item to the cart, and then write the updated cart back to the key-value store. Update the shopper's last-active timestamp in the key-value store as well.

Whenever a shopper places an order, clear any current tracking of their cart.

Every 30 seconds, iterate through the entire contents of our key-value store, looking for abandoned carts.

If a shopper's last active key is more than 30 minutes ago, we have an abandoned shopping cart.

Sends a "Shopper abandons cart" event to the derived-events-ch05 stream. The event was generated using shopper and cart information held in the key-value store.

```
        resetShopper(shopper);
      }
    }
  }
}

private static String asTimestampKey(String shopper) {
  return shopper + "-ts";
}

private static boolean isTimestampKey(String key) {
  return key.endsWith("-ts");
}

private static String extractShopper(String key) {
  return key.substring(0, key.lastIndexOf('-'));
}

private static String asCartKey(String shopper) {
  return shopper + "-cart";
}

private void resetShopper(String shopper) {
  store.delete(asTimestampKey(shopper));
  store.delete(asCartKey(shopper));
}
}
```

**Helper to reverse out a shopper's ID from a key in the key-value store**

There's a lot to take in with this Samza job: let's briefly review how our new job's `process()` and `window()` functions work. To start with `process()`:

- We are interested in *only Shopper places order* and *Shopper adds item to cart* events.
- When a *Shopper adds item to cart*, we update a copy of their cart stored in our key-value store and update our shopper's last active timestamp.
- When a *Shopper places order*, we delete all state about our shopper from the key-value store.

Our `process()` function is responsible for keeping a copy of each shopper's cart up-to-date based on their add-to-cart events; it is also responsible for understanding how recently the user added something to their cart.

Now let's briefly recap the `window()` function:

- Every 30 seconds, we scan the whole key-value store, looking for shoppers who were *last* active more than 30 minutes ago.
- We generate a *Shopper abandons cart* event for each shopper we find, detailing the contents of their shopping cart as recorded in our key-value store.
- We send each *Shopper abandons cart* event to our outbound Kafka stream.
- We delete from the key-value all values for the shoppers who just abandoned their carts.

With the code complete, let's now compile and package our new Samza job. Still from the project root, run the following:

```
$ mvn clean package
...
[INFO] Building tar: .../nile-carts/target/nile-carts-0.14.0-dist.tar.gz
[INFO] --------------------------------------------------------------------
  --
[INFO] BUILD SUCCESS
...
```

Great—everything compiles, and we have now packaged our first Samza job. We are ready to run our job on a resource management framework called Apache Hadoop YARN. We'll cover this in the next section.

## 5.5    *Running our Samza job*

Although Samza now supports being embedded into a regular JVM application (like Kafka Streams), the more common way of running a Samza job is via YARN. Unless you have previously worked with Hadoop, it's unlikely that you will have encountered YARN before—so we will introduce YARN briefly before getting our job running on it.

### 5.5.1    *Introducing YARN*

*YARN* stands for *Yet Another Resource Negotiator*—an uncomplimentary backronym for an important piece of technology. YARN is a software system that evolved out of Hadoop 1. The biggest difference between Hadoop 1 and Hadoop 2 is the separation of the cluster management responsibility into the YARN subproject.

YARN is deployed onto a Hadoop cluster to allocate resources to YARN-aware applications effectively. It has three core components:

- *ResourceManager*—The central "brain" that tracks servers in the Hadoop cluster and jobs running on those servers, and allocates compute resources to these jobs
- *NodeManager*—Runs on every server in the Hadoop cluster, monitoring the jobs and reporting back to the ResourceManager
- *ApplicationMaster*—Runs alongside each application and negotiates the required resources from the ResourceManager, and works with the NodeManager to execute and monitor each task

YARN is somewhat unfashionable these days—the Kubernetes project gets much more attention—but it is tried and tested technology (rather like ZooKeeper is), and it is widely deployed, given the pervasiveness of Hadoop 2 environments. YARN was also designed in a generic-enough way that various stream processing frameworks, completely unrelated to Hadoop, have been able to use it as their job scheduler. Samza falls into this bucket.

That's enough theory; now we need to install YARN. Fortunately for us, the Hello World project for Samza comes with a script called grid that helps you to set up YARN, as well as Kafka and ZooKeeper. This script will also check out the correct version of

Samza and build it. You worked with Kafka and ZooKeeper in the preceding chapter—but grid will set up these too if they are not still running. All the new software will be added into a subdirectory called deploy inside the root folder.

From the project root, run the grid script like so:

```
$ bin/grid bootstrap
Bootstrapping the system...
EXECUTING: stop kafka
...
kafka has started
```

You can now navigate to the YARN UI in your web browser:

```
http://localhost:8088
```

The list of All Applications will be empty. We will change this in the next section, by submitting our Samza job to YARN.

### 5.5.2   Submitting our job

Make sure that you are still in the root folder of your project, and then run the following:

```
$ mkdir -p deploy/samza
$ tar -xvf ./target/nile-carts-0.14.0-dist.tar.gz -C deploy/samza
```

We can now submit our Samza job to YARN:

```
$ deploy/samza/bin/run-app.sh \
  --config-factory=org.apache.samza.config.factories.PropertiesConfigFactory \
  --config-path=file://$PWD/deploy/samza/config/nile-carts.properties
...
2018-10-15 17:25:30.434 [main] JobRunner [INFO] job started successfully
 - Running
2018-10-15 17:25:30.434 [main] JobRunner [INFO] exiting
```

If you return to the YARN UI, you should be able to see your Samza jobs running, as in figure 5.10:

```
http://localhost:8088
```

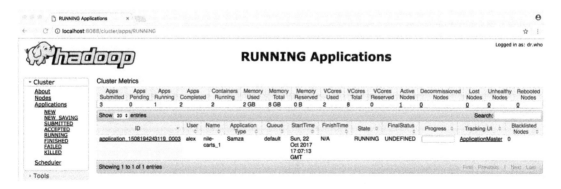

Figure 5.10   The YARN UI is now showing our Samza job in RUNNING state.

Now let's put our new job through its paces.

### 5.5.3    *Testing our job*

To test our job, we will send e-commerce events into a new Kafka topic, raw-events-ch05, and then look for events written by our Samza job into a new stream, this time called `derived-events-ch05`. If our Samza job is working, this output stream will receive fully formed *subject-verb-object* events.

Let's start by *tailing* our output stream. From the root of the project folder, run this command:

```
$ deploy/kafka/bin/kafka-console-consumer.sh \
    --topic derived-events-ch05 --from-beginning \
    --bootstrap-server localhost:9092
```

In a separate terminal, let's start a script that lets us send events into our Kafka topic for incoming events, which we will call raw-events-ch05. Samza has automatically created this Kafka topic for us, thanks to its mention in the Samza job configuration. Start up the event producer like so:

```
$ deploy/kafka/bin/kafka-console-producer.sh --topic raw-events-ch05 \
    --broker-list localhost:9092
```

Let's start by sending in a couple of add-to-basket events from different shoppers:

```
{ "subject": { "shopper": "123" }, "verb": "add", "indirectObject":
 "cart", "directObject": { "item": { "product": "aabattery", "quantity":
 12 } }, "context": { "timestamp": "2018-10-25T11:56:00" } }

{ "subject": { "shopper": "456" }, "verb": "add", "indirectObject":
 "cart", "directObject": { "item": { "product": "macbook", "quantity":
 1 } }, "context": { "timestamp": "2018-10-25T11:56:12" } }
```

Now here's the thing: by the time you run these commands, these add-to-basket events will be long in the past—far more than 30 minutes ago. If you switch back to your derived-events-ch05 stream, you should see your first generated events very soon:

```
{"subject":{"shopper":"123"},"verb":"abandon","context":{"timestamp":
 "2018-10-25T11:56:00"},"direct-object":{"cart":{"items":
 [{"product":"aabattery","quantity":12}]}}}
{"subject":{"shopper":"456"},"verb":"abandon","context":{"timestamp":
 "2018-10-25T11:56:12"},"direct-object":{"cart":{"items":
 [{"product":"macbook","quantity":1}]}}}
```

So far, so good; next let's try an add-to-basket followed quickly by a checkout event:

```
{ "subject": { "shopper": "789" }, "verb": "add", "indirectObject":
 "cart", "directObject": { "item": { "product": "skyfall", "quantity":
 1 } }, "context": { "timestamp": "timestamp":"2018-10-25T12:00:00" } }

{ "subject": { "shopper": "789" }, "verb": "place", "directObject": {
 "order": { "id": "123", "value": 12.99, "items": [ { "product":
```

"skyfall", "quantity": 1 } ] } }, "context": { "timestamp":
"2018-10-25T12:01:00" } }

Assuming you managed to send in both events within the same 30-second window, you should expect no *Shopper abandons cart* event in `derived-events-ch05`. This is because shopper `789` placed their order within the allowed time.

And that's it: our Samza job is monitoring a stream of incoming events to detect abandoned shopping carts. You can try it out with a few more events of your own creation—just be careful to follow the existing schema.

From the project root, you can now run the following command in order to stop all processes:

```
$ bin/grid stop all
EXECUTING: stop all
EXECUTING: stop kafka
EXECUTING: stop yarn
stopping resourcemanager
stopping nodemanager
EXECUTING: stop zookeeper
```

### 5.5.4 *Improving our job*

This Samza job is a good first stab at an abandoned cart detector. If you have time, you could enhance and extend it in a variety of ways:

- De-duplicate items in the cart recorded in Samza's key-value store. If a shopper adds one copy of the *Skyfall* DVD to their basket twice, this is rationalized to one item with a quantity of 2.
- Use the timestamps of incoming events (from the event context) to determine when the shopper was last active, rather than basing it on the time when `process()` is running, as now.
- Define a new *Shopper removes item from cart* event, and use these events to remove items from the shopper's cart as stored in Samza's key-value store.
- Base each shopper's last-active timestamp on all events for that shopper, not just *Shopper adds item to cart* events.
- Explore more-sophisticated ways of determining whether a cart is abandoned, instead of the strict 30-minute cutoff. You could try varying the cutoff based on the shopper's previously observed behaviors.

## *Summary*

- Processing multiple events from a stream requires state. This state allows the app to "remember" important attributes about individual events, across many events.
- State for event stream processing can be kept in-memory (transient), stored locally on the processing instance, or written to a remote database. A stream processing framework can help us to manage this process.

- Stream processing frameworks also help us to apply time windows to our streams, have delivery guarantees for our events, distribute our stream processing across multiple servers, and tolerate failures in those individual servers and processes.
- Popular stream-processing frameworks include Apache Storm, Apache Samza, Spark Streaming, Kafka Streams, and Apache Flink.
- Samza is a stateful stream-processing framework with a relatively low-level API, making it the "Hadoop MapReduce of stream processing."
- Samza's process() function lets us update values in our key-value store every time we get a new event from our incoming event stream in Kafka.
- Samza's window() function lets us regularly review what has changed in our application.
- By clever data modeling in our key-value store (for example, with composite keys), we can detect sophisticated patterns and behaviors, such as shopping-cart abandonment.

# Part 2

# Data engineering with streams

In part 2 of this book, we'll explain how to describe, store, and archive streams of events. We'll also introduce railway-oriented processing as a way to properly design for failures in stream processing architectures.

# Schemas

**This chapter covers**

- Event schemas and schema technologies
- Representing events in Apache Avro
- Self-describing events
- Schema registries

In the first part of this book, we took a wide-ranging look at event streams and the unified log, using fictitious online retailer Nile. We looked in depth at adding a unified log to our organization and experimented with different stream-processing frameworks to work with the events in our Kafka topics.

But like fast-food addicts, we didn't spend a lot of time thinking about the *quality* of the events that we were feeding into our unified log. This part of the book aims to change this, by looking much more closely at the way we model the events flowing through our unified log, using *schemas*.

Working for Plum, a fictitious global consumer-electronics manufacturer, we will introduce Plum's first event, a regular health-check "ping" emitted from each NCX-10 machine on the factory floor. Like every unified log, Plum's is fundamentally decoupled: consumers and producers of event streams have no particular knowledge of each other. This puts the onus on event schemas to serve as the contract between Plum's event consumers and producers.

How should we define Plum's event schemas? The chapter introduces four widely used schema technologies: Apache Avro, JSON Schema, Apache Thrift, and Google's protocol buffers. More than just data serialization systems, schema technologies like Avro offer schema evolution support, generation of bindings for your programming language, and multiple options for encoding the events on disk.

Adopting Avro for Plum, we will model the NCX-10 health-check event in the Avro schema language, and then autogenerate Plain Old Java Object (POJO) bindings for that event in Java. We'll then test this with a simple Java app that deserializes a JSON-format Avro health-check event, prints it out as a POJO, and then reserializes it into Avro's binary format.

With simple Avro parsing under our belt, we'll continue into design questions around associating events with their schemas. We suggest a couple of possible approaches to making this event-schema association, before arguing strongly for *self-describing* events. These are events that include a metadata "envelope," referencing the schema, alongside the event itself. To work with these events, you first read the event's outer envelope to discover the event's Avro schema, and then retrieve the schema and use it to deserialize the event's data payload itself.

Before we know it, we will be proliferating schemas for our employer Plum. And these have to live somewhere—namely, in a *schema registry*. We will wrap up the chapter with a brief look at the core attributes of a schema registry, before introducing the two most widely used schema registries: Confluent Schema Registry and Snowplow's own Iglu schema registry. These two registries are broadly similar but have interesting differences of design decision, which we'll look at last.

Let's get started!

## 6.1 *An introduction to schemas*

To start working with schemas, we'll first need some events. For this part of the book, we'll leave Nile behind and work with another fictitious company with a unified log, called Plum. Let's introduce Plum and its event streams.

### 6.1.1 *Introducing Plum*

Let's imagine that we are in the Business Intelligence (BI) team at Plum, a global consumer-electronics manufacturer. Plum is in the process of implementing a unified log. For unimportant reasons, its unified log is a hybrid, using Amazon Kinesis for certain streams and Apache Kafka elsewhere; Amazon Kinesis (https://aws.amazon.com/kinesis/) is a hosted unified log service available as part of Amazon Web Services. In reality, using both Kafka and Kinesis is a fairly unusual setup, but it has the advantage for us that we can work with both Kinesis and Kafka in this section of the book.

At the heart of the Plum production line is the NCX-10 machine, which stamps new widgets out of a single block of steel. Plum has 1,000 of these machines in each of its 10 factories. Our corporate overlords at Plum want us to program each machine to emit key metrics in the form of a health check to a Kinesis stream every second, as shown in figure 6.1. A Kinesis stream is similar to a Kafka topic (don't worry, we'll explore Kinesis in detail in the next chapter).

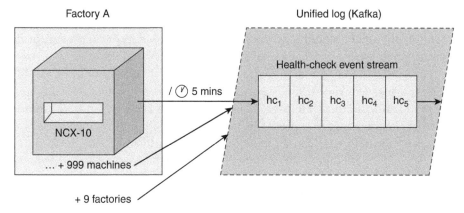

**Figure 6.1  All of the NCX-10 machines in Plum's factories should emit a standard health-check event to a Kinesis stream every second.**

Across Plum's 10 factories, we have 10,000 machines. With each machine emitting a health check every second, that's 36 million health checks landing in the Kinesis stream each hour, a substantial first event source for Plum's unified log.

We sit down with the plant maintenance team at Plum to find out what health-check information we can retrieve from the NCX-10 machines. We discover a few interesting data points:

- The name of the factory that the machine is installed in.
- The machine's serial number, which is a string.
- The machine's current status, which can be one of STARTING, RUNNING, or SHUTTING_DOWN.
- The time when the machine was last started, which is a Unix timestamp accurate to milliseconds.
- The machine's current temperature, in Celsius.
- Whether or not the machine has been earmarked as being end-of-life, meaning that it should be scrapped soon.
- If the factory has multiple floors, the machine knows which floor it is situated on.

In part 1 of this book, we used JSON as a way of encoding all of Nile's online shopping-related events. Channeling this expertise for Plum, we can quickly sketch out an example JSON instance representing a hypothetical health check from one of the NCX-10 machines:

```
{ "factory": "Factory A",
  "serialNumber": "EU3571",
  "status": "RUNNING",
  "lastStartedAt": 1539598697944,
```

```
"temperature": 34.56,
"endOfLife": false,
"floorNumber": 2
  }
```

The plant engineers squint at the preceding example JSON and confirm that this is pretty much everything that an NCX-10 machine can usefully emit.

This JSON is a great start. If this were still part 1 of the book, we would take this JSON format and run with it, designing and building stream processing applications to read and write these JSON-based events. But this is part 2, and for Plum we can go further: we are going to replace this JSON format with a formal *schema* for our health-check event. Before we can do this, you need to understand exactly what a schema is, and why it is so useful.

### 6.1.2   *Event schemas as contracts*

Remember that the powers-that-be at Plum have so far given us a narrow brief: programming each of the NCX-10 machines to emit a regular health-check event. We don't yet know how Plum wants to use this event, or even which teams within Plum will be tasked with working with these events in Plum's unified log.

This is a common occurrence: in the real world, someone else, maybe even a different team or department or country, will most often consume the event stream that you create. A unified log is a fundamentally *decoupled* architecture: consumers and producers of event streams have no particular knowledge of each other. This is in stark contrast to a traditional single monolithic software project, where you can see all parts of the code evolving in lockstep within source control, with a compiler and test suite constantly enforcing the integrity of the whole system.[1]

If you think about it, the decoupled unified log approach is analogous to the widgets that Plum's beloved NCX-10 machines are churning out. Plum doesn't necessarily know which companies will buy the widgets, or what purposes those customers will put the widgets to. How does Plum guarantee that its widgets are fit-for-purpose for its customers? They do this by signing contracts with their customers that dictate certain standards that their widgets will conform to. Perhaps the widgets will always be SAE steel grade 12L14, pass the GS quality and safety test, and use the Unified Thread Standard for its screw threads. With these standards enforced in contracts, Plum's customers can be confident that they can build products that make use of Plum's widgets.

The consumers of the event streams within our unified log need exactly the same kind of guarantees as the buyers of Plum's widgets. We can give them these guarantees by using *schemas* for the events we are storing in our unified log. *Schema* is the Greek word for *shape*, and an event schema is simply that: a declaration that some set of events in our unified log will follow a predefined shape.

---

[1] You can find more information about continuous integration at Wikipedia: https://en.wikipedia.org/wiki/Continuous_integration.

In the absence of formal integrations between systems, our event schemas are the closest we come to a *contract* between the system generating an event stream and the systems consuming that stream. Figure 6.2 illustrates this contract.

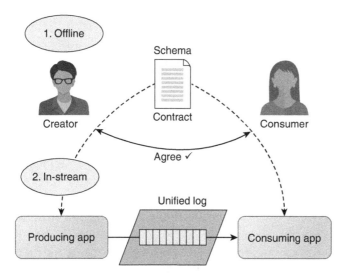

**Figure 6.2**  **The producer of the event agrees to the schema of the event with the initial known consumer of the event. This acts as a contract between both parties. Then, in the stream, the producing app creates events using the agreed-upon schema, and the consuming app can happily read those events, safe in the knowledge that the events will conform to the schema.**

By agreeing to the schema up-front and writing our code in a way that ensures that the events we generate conform to that schema, we can avoid a huge amount of pain for whoever is consuming those events later. But how do we represent our event schemas? The good news is that we have a choice of a variety of schema technologies, often called *data serialization systems*, which have emerged over the past few years. Before we introduce these, let's briefly cover the capabilities of these technologies.

### 6.1.3 *Capabilities of schema technologies*

In part 1, we built applications that produced and consumed events that were represented using plain JSON. The events we created for online retailer Nile certainly each had a *shape*, but that shape was not formally documented in a machine-readable way; we could say that Nile's events had *implicit* schemas rather than explicit.

To be clear, JSON is a data serialization system, but it's not one that allows us to provide the kinds of contracts that we need for our unified log. The schema technologies that we will look at in the next section all provide a *schema language* to precisely define the data types of the individual properties of your business entities.

If you have ever worked with a *strongly typed* programming language, you will be familiar with data types. Simple data types such as string or integer are almost always well represented by these schema languages; different schema technologies will, however, have varying support for the following:

- *Compound data types* such as arrays and records or objects
- *Less common data types* such as timestamps, UUIDs, and geographical coordinates

Going further, the schema technologies we will look at all additionally exhibit at least some of these six capabilities (the abbreviations in parentheses refer to table 6.1 in the following subsection):

- *Multiple schema languages* (LNG)—Some schema technologies provide multiple ways in which you can express the data types for your business entities. For example, you may be able to write your schemas declaratively in JSON, or there may some form of interface description language (IDL) with a more C-like or Java-like syntax.
- *Validation rules* (VLD)—Some schema technologies go further than data types, by letting you express validation rules (sometimes called *contracts*) on the event's properties. For example, you might express that a longitude is not just a floating-point number, but a floating-point number that can't be more than 180 or less than –180, and that latitude must be no more than 90 or less than –90.
- *Code generation* (GEN)—Whatever syntax the schema is expressed in, we are likely to want to also interact with events represented by the schema in code we write (for example, our stream processing apps). To facilitate this, schema technologies often support code generation, which will generate idiomatic classes or records in your preferred language from the schema.
- *Multiple encodings* (ENC)—Some schema technologies support multiple encodings of the data, often a compact binary format and a human-readable format (perhaps JSON-based).
- *Schema evolution* (EVO)—The properties in our Plum events will likely evolve over time. Some of the more sophisticated schema technologies have built-in support for schema evolution, which makes it easier to consume different versions of the same schema.
- *Remote procedure calls* (RPC)—This is not a feature that we need, but some data serialization systems also come with a mechanism for building remote procedure calls, distributed functions that use data types for expressing the functions' arguments and return values.

Don't worry if some of these capabilities sound a little abstract right now. The rest of this chapter will use Plum's business requirements to make all this much more concrete.

### 6.1.4  Some schema technologies

A variety of schema technologies, also known as *data serialization systems,* have emerged over the past few years, each with subtly different capabilities. Table 6.1 introduces four of the most widely used systems and lays out their design relative to the six capabilities introduced in the preceding section.

**Table 6.1  Examples of schema technologies**

| Schema tech | LNG | VLD | GEN | ENC | EVO | RPC |
|---|---|---|---|---|---|---|
| Apache Avro | JSON, IDL | No | Yes | Binary, JSON | Yes | Yes |
| Apache Thrift | IDL | No | Yes | Five encodings | Yes | Yes |
| JSON Schema | JSON | Yes | No | JSON | No | No |
| Protocol buffers | IDL | No | Yes | Binary, JSON | No | Yes (gRPC) |

Let's go through each of these schema technologies briefly.

**APACHE AVRO**

*Apache Avro* (https://avro.apache.org/) is an RPC and data serialization system that was developed as part of the Apache Hadoop project, and shares many of the same authors, including Hadoop pioneer Doug Cutting.

Avro has a declarative JSON-based schema language for describing data types, as well as an alternative language, called Avro IDL, which is more C-like. Avro has two representations for data: a compact binary encoding and a human-readable JSON encoding; the latter follows a few additional rules to represent Avro features that are not natively supported in JSON. The binary encoding is used much more widely than the JSON encoding, and as such the tooling for the binary encoding tends to be more fully featured.

As a software engineer, you can interact with Avro in one of two ways. The first is to use code generation ahead of time, creating representations for the Avro data types in your preferred programming language; you can then round-trip the Avro-encoded data to classes or records in your code, and back again. The second approach involves using the schema at runtime to parse the data in a generic way; this is a good fit for dynamic languages and for situations (common in a unified log) in which the record types to process are not known ahead of time.

With its origins in the Hadoop project, Avro was designed from the outset for use in data processing, and as a result has a sophisticated take on schema evolution. When a consuming application is reading an Avro record, it must have a copy of the schema that was used to write the record, but it can also supply another version of that same schema; these are the *writer's schema* and the *reader's schema,* respectively. Avro will then transparently use resolution rules between the two schema versions to represent the data in the reader's schema. This allows an application to happily process an archive of events written by multiple historic versions of a given schema.

### APACHE THRIFT

*Apache Thrift* (https://thrift.apache.org/) is pitched as a framework for cross-language services development. As part of this, it includes a definition language that allows you to define data types as well as service interfaces. One of the biggest selling points is its broad language support: Thrift has first-class support for C++, Java, Python, PHP, Ruby, Go, Erlang, Perl, Haskell, C#, Cocoa, JavaScript, Node.js, Smalltalk, OCaml, and Delphi.

Unlike the other schema technologies, Thrift is much less opinionated about how the data should be encoded: it offers three binary encodings and two JSON-based encodings. When you define a schema in Thrift, you manually assign each property a tag (1, 2, 3, and so forth), and these tags are stored in the encoded data along with the data type; this is how Thrift supports schema evolution: an unknown property can be skipped, or a renamed property correctly identified.

Thrift was created at Facebook. The story goes that "Xooglers" created it because they missed using Google's own protocol buffers, which hadn't been open sourced at that point.

### JSON SCHEMA

*JSON Schema* (https://json-schema.org/) is slightly different from the other schema technologies described here. JSON Schema is a declarative language, itself written in JSON, for describing your JSON data's format; it is easily written by humans and makes it possible for computers to validate and parse individual JSON data. If you are familiar with XML, the relationship between JSON and JSON Schema is not dissimilar to that between XML and its document type definitions (DTDs).

JSON Schema doesn't concern itself with data serialization at all; it's simply an overlay for defining the shape of JSON data. It also doesn't depend on (or offer) code generation, although there are community-led initiatives to implement code generation for various languages.

Although not as fully featured as Avro, JSON Schema does have two distinct strengths:

- *Rich validation rules*—It's possible to express sophisticated data validation rules in JSON Schema, including minima and maxima for numbers, and regular expressions for strings. Used creatively, this allows you to define your own simple data types.
- *It's just JSON*—Because JSON Schema is simply a schema overlay over plain JSON, you don't have to rewrite your systems to use another data serialization system. If you are comfortable with JSON, JSON Schema is quick to pick up. You can also backfill schemas for existing JSON data after the fact by using tools such as Schema Guru (https://github.com/snowplow/schema-guru).

### PROTOCOL BUFFERS

*Protocol buffers* (https://developers.google.com/protocol-buffers/) are a schema technology and data serialization mechanism from Google, currently on its third major version (and open sourced from its second major version). Protocol buffers are similar to Thrift; they support a protocol definition syntax that lets the user define data structures in .proto files.

Data in protocol buffers is serialized into a binary format; there is also a self-describing ASCII format, but this does not support schema evolution. As with Thrift, integers are used to tag each property within the data structure and to handle schema evolution. One neat aspect of protocol buffers is that arrays are represented with the `repeated` modifier, and the same encoding is used for repeated, optional, and required properties, making it possible to easily migrate a property from being, for example, optional to being an array.

Protocol buffers can be used as a standalone schema technology, but they are also closely associated with and used with Google's RPC framework, called gRPC.

This concludes our whirlwind introduction to four major schema technologies—but how do we choose between these for our bosses at Plum?

### 6.1.5 *Choosing a schema technology for Plum*

Choosing our schema technology is one of the most important decisions we will make for our unified log at Plum. We can adopt different stream-processing frameworks, but we get to adopt only one schema technology. All of the events we write into our unified log will be stored with these schemas; the archive of events from Plum's unified log (which hopefully will stretch into many years) will be represented in this format too.

Avro, JSON Schema, and protocol buffers are all growing in popularity; interest in Thrift is quite possibly growing too. So how do we choose between them? Here are simple rules that might help:

- If you are using or plan to use gRPC, consider using protocol buffers for your unified log.
- Likewise, if you are using Thrift RPC already, consider using Thrift.
- If you have existing batch- or stream-processing systems that make heavy use of JSON, or you expect a lot of event authoring to be done by developers who prefer JSON, consider JSON Schema.
- Otherwise, use Avro.

In the case of Plum, none of the first three rules are met, so we are going to go with Avro. Enough theory—let's get started.

## 6.2 *Modeling our event in Avro*

From our colleagues on the factory floor, we know what data points our health-check events need to contain, and from the brief review in the previous section, we know that we want to use Avro as our schema technology at Plum. So, we are ready now to model our health-check event in Avro.

### 6.2.1  *Setting up a development harness*

Avro schemas can be defined in plain JSON or Avro IDL text files, but the schema file needs to live somewhere, so we'll create a simple Java app to hold it, called `SchemaApp`. This will give us a harness to experiment with Avro's code-generation capabilities, as well as to explore its encodings.

We are going to write the harness app by using Gradle. First, create a directory called plum, and then switch to that directory and run the following:

```
$ gradle init --type java-library
...
BUILD SUCCESSFUL
...
```

Gradle has created a skeleton project in that directory, containing a couple of Java source files for stub classes called Library.java and LibraryTest.java. Delete these two files, because we will be writing our own code shortly.

Next let's prepare our Gradle project build file. Edit the file build.gradle and replace its current contents with the following listing.

---

**Listing 6.1  build.gradle**

```
plugins {
  id "com.commercehub.gradle.plugin.avro" version "0.9.1"    ◁─┐  A Gradle plugin to
}                                                               generate Java code from
                                                                Apache Avro schemas
apply plugin: 'java'
apply plugin: 'application'

sourceCompatibility = '1.8'

mainClassName = 'plum.SchemaApp'

repositories {
  mavenCentral()
}

version = '0.1.0'

dependencies {
  compile 'org.apache.avro:avro:1.8.2'    ◁─┐  The latest release
}                                             of Avro at the time
                                              of writing
jar {
  manifest {
    attributes 'Main-Class': mainClassName
  }

  from {
    configurations.compile.collect {
      it.isDirectory() ? it : zipTree(it)
```

```
    }
  } {
    exclude "META-INF/*.SF"
    exclude "META-INF/*.DSA"
    exclude "META-INF/*.RSA"
  }
}
```

Let's confirm that we can build this Gradle project:

```
$ gradle compileJava
...
BUILD SUCCESSFUL
...
```

Good—now we are ready to work on the schema for our health-check event.

### 6.2.2 *Writing our health check event schema*

To add our new Avro schema for the NCX-10 machine's health-check event, create a file at this path:

```
src/main/resources/avro/check.avsc
```

Populate this file with the Avro schema in the following listing. Note that we are using Avro's JSON-based schema syntax rather the Avro IDL to model this event; this is what the .avsc file extension signifies.

```
Listing 6.2   check.avsc
```

```
{ "name": "Check",
  "namespace": "plum.avro",
  "type": "record",
  "fields": [
    { "name": "factory", "type": "string" },
    { "name": "serialNumber", "type": "string" },
    { "name": "status",
      "type": {
        "type": "enum", "namespace": "plum.avro", "name": "StatusEnum",
        "symbols": ["STARTING", "RUNNING", "SHUTTING_DOWN"]
      }
    },
    { "name": "lastStartedAt", "type": "long",
      "logicalType": "timestamp-millis" },
    { "name": "temperature", "type": "float" },
    { "name": "endOfLife", "type": "boolean" },
    { "name": "floorNumber", "type": ["null", "int"] }
  ]
}
```

This is a densely packed Avro schema file. Let's break it down: the top-level entity is a record called Check, which belongs in the plum.avro namespace (as would all of our entities). Our Check record contains seven fields, which correspond to the seven data points identified for a health-check event:

- The name of the factory in which the machine is installed is of type string.
- The machine's serial number is another string.
- The machine's current status is an Avro enum (short for *enumeration*), which can be one of STARTING, RUNNING, or SHUTTING_DOWN. An enum is a *complex type*, so we need to provide it with a namespace (plum.avro) and a name (StatusEnum).
- The time when the machine was last started, which is a Unix timestamp accurate to milliseconds. This is stored as a long, but Avro also lets us specify a *logical type* for the field, which gives a hint as to how a parser should handle the underlying type.
- The machine's current temperature, in Celsius, is a float.
- Whether or not the machine has been earmarked as being end-of-life is a boolean.
- The floor number is stored as a *union type* of an int or a null. If the factory does not have multiple floors, this field will be set to null. Otherwise, it will be an integer.

A minor piece of housekeeping—we need to soft-link the resources/avro subfolder to another location so that the Avro plugin for Gradle can find it:

```
$ ln -sr src/main/resources/avro src/main
```

With our Avro schema defined, let's use the Avro plugin in our build.gradle file to automatically generate Java bindings for our schema:

```
$ gradle generateAvroJava
:generateAvroProtocol UP-TO-DATE
:generateAvroJava

BUILD SUCCESSFUL

Total time: 8.234 secs
...
```

You will find the generated files inside your Gradle build folder:

```
$ ls build/generated-main-avro-java/plum/avro/
Check.java  StatusEnum.java
```

These files are too lengthy to reproduce here, but if you open them in your text editor, you will see that the files contain POJOs to represent the one record and the one enumeration that make up our health-check event.

So far, so good: we now have a model in Java for our Avro health check-event. In the next section, we will write simple Java code to work with these health checks.

### 6.2.3 *From Avro to Java, and back again*

Remember that Avro has two representations, a human-readable JSON encoding and a more efficient binary encoding. In this section, we will round-trip a health-check event from Avro's JSON-based encoding into a Java object, and then back again into Avro's binary encoding. This is not particularly useful for Plum, but it will give you a chance to become familiar with Avro's two representations and show you how to interact with Avro from a regular programming language such as Java.

Add the following code into a new file called src/main/java/plum/AvroParser.java.

**Listing 6.3 AvroParser.java**

```java
package plum;

import java.io.*;
import java.util.*;
import java.util.Base64.Encoder;

import org.apache.avro.*;
import org.apache.avro.io.*;
import org.apache.avro.generic.GenericData;
import org.apache.avro.specific.*;

import plum.avro.Check;                         ◁──── Import the Check POJO
                                                      autogenerated from the
                                                      bundled schema.
public class AvroParser {

  private static Schema schema;                       Statically initialize this
  static {                                            Schema object from the
    try {                              ◁──────         bundled schema.
      schema = new Schema.Parser()
        .parse(AvroParser.class.getResourceAsStream("/avro/check.avsc"));
    } catch (IOException ioe) {
      throw new ExceptionInInitializerError(ioe);
    }
  }

  private static Encoder base64 = Base64.getEncoder();

  public static Optional<Check> fromJsonAvro(String event) {

    InputStream is = new ByteArrayInputStream(event.getBytes());
    DataInputStream din = new DataInputStream(is);

    try {
      Decoder decoder = DecoderFactory.get().jsonDecoder(schema, din);
      DatumReader<Check> reader = new SpecificDatumReader<Check>(schema);
      return Optional.of(reader.read(null, decoder));
    } catch (IOException | AvroTypeException e) {
      System.out.println("Error deserializing:" + e.getMessage());
      return Optional.empty();
    }
  }
}
```

**Deserialize our event into a Check POJO, boxed in an Optional.**

```
    public static Optional<String> toBase64(Check check) {

      ByteArrayOutputStream bout = new ByteArrayOutputStream();

      DatumWriter<Check> writer = new SpecificDatumWriter<Check>(schema);
      BinaryEncoder encoder = EncoderFactory.get().binaryEncoder(bout, null);
      try {
        writer.write(check, encoder);
        encoder.flush();
        return Optional.of(base64.encodeToString(bout.toByteArray()));    ◄─┐
      } catch (IOException e) {                                             │
        System.out.println("Error serializing:" + e.getMessage());         │
        return Optional.empty();                                           │
      }                                   Serialize our event in its binary │
    }                                     format, and then Base64 encode it.│
  }
}
```

The AvroParser file is simple; it consists of three parts:

- *Initialization code*—This gives us a reusable representation in Java of the Avro health-check schema; it also gives us a Base64 encoder, which we'll use later.
- A *static function*, `fromJsonAvro`—This converts an incoming event, which is hopefully a health check in Avro's JSON format, into a `Check` POJO. We box the return value in a Java 8 `Optional` to cover the case where the health check couldn't be correctly deserialized.
- A *static function*, `toBase64`—This converts our Check POJO into an Avro binary record, and then Base64 encodes that byte array to make it more human-readable. Again, we box the return value in an `Optional` to cover any serialization problems.

We can now stitch these two functions together via a new `SchemaApp` class containing our `main` method. Create a new file called src/main/java/plum/SchemaApp.java and populate it with the contents of the following listing.

**Listing 6.4   SchemaApp.java**

```
package plum;

import java.util.*;

import plum.avro.Check;

public class SchemaApp {
                                                         Attempt to    ┐
  public static void main(String[] args){           deserialize our event │
    String event = args[0];                            into a Check POJO.  │
                                                                           │
    Optional<Check> maybeCheck = AvroParser.fromJsonAvro(event);    ◄──────┘

    maybeCheck.ifPresent(check -> {               ◄──┐  Proceed only if we
      System.out.println("Deserialized check event:");  │  have a Check.
      System.out.println(check);
```

```
                    Optional<String> maybeBase64 = AvroParser.toBase64(check);
                    maybeBase64.ifPresent(base64 -> {
                        System.out.println("Re-serialized check event in Base64:");
                        System.out.println(base64);
                    });
                });

        }
    }
```

**Proceed only if we have a byte string.**

**Attempt to serialize our Check into a Base64-encoded byte string.**

We will pass a single argument into our `SchemaApp` on the command line: a string hopefully containing a valid NCX-10 health-check event, in Avro's JSON format. Our code then attempts to deserialize this string into a `Check` POJO. If this succeeds, we proceed to print out the `Check` and then convert it back into Avro's binary representation, Base64-encoded so that we can print it out easily.

Let's build our app now. From the project root, the plum folder, run this:

```
$ gradle jar
...
BUILD SUCCESSFUL

Total time: 25.532 secs
```

Great—we are now ready to test our new Avro-powered schema app.

### 6.2.4 Testing

Remember back in section 6.1.1, where we introduced Plum and our NCX-10 health-check event, conveniently represented in JSON? Almost the same JSON is valid as an Avro representation of the same event. Here is the Avro representation:

```
{ "factory": "Factory A",
  "serialNumber": "EU3571",
  "status": "RUNNING",
  "lastStartedAt": 1539598697944,
  "temperature": 34.56,
  "endOfLife": false,
  "floorNumber": { "int": 2 }
}
```

The only difference is the `{"int": ... }` syntax boxing the `floorNumber` property's value of 2. Avro requires this because the floor number in our health-check event is optional, and Avro represents optional fields as a *union* of null and the other type (in this case, an integer). We need to use the `{"int": ... }` syntax to tell Avro that we want to treat the value of this union type as an integer, rather than a null.

Let's take this JSON representation of our health-check event and pass it into our freshly built Java app:

```
$ java -jar ./build/libs/plum-0.1.0.jar "{\"factory\":\"Factory A\",
 \"serialNumber\":\"EU3571\",\"status\":\"RUNNING\",\"lastStartedAt\":
 1539598697944,\"temperature\":34.56,\"endOfLife\":false,
```

```
\"floorNumber\":{\"int\":2}}"
Deserialized check event:
{"factory": "Factory A", "serialNumber": "EU3571", "status": "RUNNING",
 "lastStartedAt": 1539598697944, "temperature": 34.56, "endOfLife": false,
 "floorNumber": 2}
Re-serialized check event in Base64:
EkZhY3RvcnkgQQxFVTM1NzEC3L2Zm+dVcT0KQgACBA==
```

Success! Our Java app has taken the incoming Avro event, successfully deserialized it into a Java POJO, printed that out, and then converted it back into a Base64-encoded binary string. Note how brief the Base64 string is. This representation is much more succinct than the JSON representation, for two reasons:

- The binary representation does not include property tags like `factory` or `status`. Instead, properties are identified positionally by using the associated schema.
- The binary representation can represent certain values efficiently—for example, using IDs for enum values, and 1 or 0 for Boolean true or false.

Before we move on, let's confirm that an invalid event fails processing:

```
$ java -jar ./build/libs/plum-0.1.0.jar "{\"factory\":\"Factory A\"}"
Error deserializing:Expected string. Got END_OBJECT
```

Good: the event correctly failed to deserialize, because various required properties are missing.

That completes our initial work with Avro for Plum. But the tight interdependency between Avro instances and their schemas raises interesting data-modeling questions, which we'll look at next.

## 6.3   *Associating events with their schemas*

Think about the preceding section: we deserialized an incoming Avro event in JSON format to a Java `Check` POJO, and then converted it back into a Base64-encoded version of that same event's binary encoding. But how did we know that the incoming Avro event was an NCX-10 health-check event? How did we know which schema to use to parse that incoming payload?

The simple answer is that we didn't. We assumed that the incoming string would be a valid NCX-10 health-check event, represented in Avro's JSON format. If that assumption was incorrect and the event did not match our Avro schema, we simply threw an error.

In this section, we will consider more-sophisticated strategies for associating events in a unified log with their schemas.

### 6.3.1   *Some modest proposals*

Let's imagine for a minute that we have implemented Apache Kafka as part of Plum's unified log. How do we write and deploy a Kafka worker that can successfully consume NCX-10 health-check events? Three potential strategies are described next.

### ONE KAFKA TOPIC PER EVENT TYPE (HOMOGENEOUS STREAMS)

In this model, Plum establishes a convention that each distinct event type will be written to its own Kafka topic. This needs to be specified down to the version of the schema and its format (binary or JSON). An example name for this specific Kafka topic might be `ncx10_health_check_v1_binary`, as illustrated in figure 6.3.

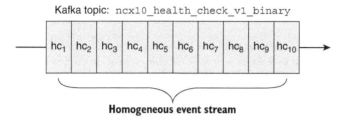

**Figure 6.3   A homogeneous Kafka topic, containing only events in the Avro binary representation of version 1 of the NCX-10 health check. A Kafka worker can consume this safely in the knowledge that all records in this topic can be deserialized using the specific Avro schema.**

The merit of this approach is its simplicity: the Kafka topic is completely homogeneous, containing only NCX-10 health checks. But as a result, we will end up proliferating event streams at Plum: one event stream (aka Kafka topic) for each version of each schema.

Therefore, if we want to do something as simple as parse multiple versions of the NCX-10 health-check event, we would have to write a stateful stream-processing app that unites data across multiple Kafka topics, one per schema version. And an analysis across all of our event types (for example, to count events per hour) is even more painful: we have to join each and every Kafka topic together in a stateful stream processor to achieve this.

Heterogeneous streams that contain multiple event types are much easier to work with. But how do we know which events they contain?

### TRIAL-AND-ERROR DESERIALIZATION (HETEROGENEOUS STREAMS)

You could call this the *brute-force* approach: Plum would mix lots of event types into one stream, and then workers would attempt to deserialize events into any given schema that they are interested in. If the event fails to deserialize into the appropriate POJO, the worker ignores that event. Figure 6.4 shows this approach.

As soon as we go beyond toy applications, this approach is incredibly inefficient. Imagine that we have an application that needs to work on five event types. Deserialization is a computationally expensive task, and our application will attempt to deserialize each event up to five times before moving on to the next event.

There has to be a better way than this brute-force approach. We cover this next.

Kafka topic: all events

**Figure 6.4  Our consuming application is interested in only two event types (***health check*** and** ***machine restart***) and uses a trial-and-error approach to identify these events in the Kafka topic. The amount of wasteful deserialization increases with the number of event types that our consuming app is concerned with.**

#### SELF-DESCRIBING EVENTS (HETEROGENEOUS OR HOMOGENEOUS STREAMS)

In this approach, we again mix many event types into a single stream. The difference this time is that each event has a simple metadata "envelope" attached to it, which tells any consuming application which schema was used to serialize the event. We can call these events *self-describing events,* because the event carries with it at all times the information about what schema this instance is associated with. Figure 6.5 depicts this approach.

Kafka topic: `self_describing_events`

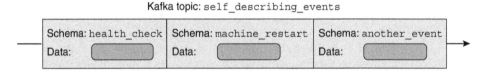

**Figure 6.5  In this Kafka topic, we can see three self-describing events: a health check, a machine restart, and another event. Each event consists of a schema describing the event, plus the event itself.**

Working with self-describing events is a two-step process:

1 Parse the event's metadata envelope to retrieve the identifier for the event's schema.
2 Parse the event's data portion against the identified schema.

This is a powerful approach: for the cost of a little more deserialization work, we now have a much more flexible way of defining our events. We can write these events into a single heterogeneous stream, but equally we can then "shred" that stream into substreams of single or associated event types. Because the event's schema travels with the event itself, we can send the event on anywhere without losing the necessary metadata to parse the event. This is visualized in figure 6.6.

An important thing to note: with self-describing events, we are always talking about adding some kind of *pointer* to the event's original schema to the event. This pointer is

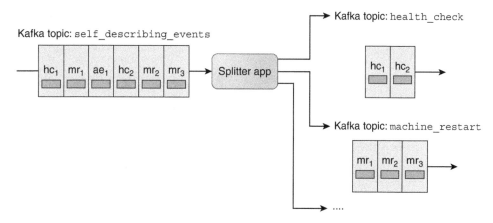

**Figure 6.6 With self-describing events, we can switch between heterogeneous streams and homogeneous streams at will, because a reference to the schema travels with the event. In this example, we have a "splitter app" separating *health checks* and *machine restarts* into dedicated Kafka topics.**

a reference rather than the schema itself, not least because the schema itself is often huge—larger even than the event payload!

Let's build a self-describing event for Plum in the next section.

### 6.3.2 A self-describing event for Plum

How should we represent a self-describing event in Apache Avro? We need a metadata "envelope" to wrap the event and record which schema was used to serialize the event. It makes sense for us to define this in Avro itself, and we've set out a simple proposal in the following listing.

**Listing 6.5 self_describing.avsc**

```
{ "name": "SelfDescribing",
  "namespace": "plum.avro",
  "type": "record",
  "fields": [
    { "name": "schema", "type": "string" },
    { "name": "data", "type": "bytes" }
  ]
}
```

You can see two fields in the `SelfDescribing` record:

- `schema` is a string to identify the schema for the given event.
- `data` is a sequence of 8-bit unsigned bytes, itself containing the Avro binary representation of the given event.

How would we identify the schema in a string format? Let's use something simple, such as this:

```
{vendor}/{name}/{version}
```

This format is a simplified implementation of the schema URIs we use in our Iglu schema registry system at Snowplow (https://github.com/snowplow/iglu). In this implementation are the following:

- vendor tells us which company (or team within the company) authored this schema. This can be helpful for telling us where to find the schema, and helps prevent naming conflicts (for example, if Plum and one of its software partners both define an ad click event).
- name is the name of the event.
- version gives the version of the given event. For simplicity at Plum, we will use an incrementing integer for the version.

In the case of the initial version of our NCX-10 health-check event, the schema string would look like this:

```
com.plum/ncx10-health-check/1
```

This self-describing envelope is a good fit for Avro's binary representation. But an interesting dilemma crops up when we think about Avro's JSON-based format. We could use the same approach, in which case a self-describing health-check event would look like this:

```
{ "schema": "com.plum/ncx10-health-check/1",
  "data": <<BYTE ARRAY>>
}
```

But there isn't a lot of point in this format: it's less compact than the binary representation, the data portion is itself still in the binary representation, and it's still not human-readable. It has little going for it! A better implementation of self-describing Avro for the JSON format would look something like this:

```
{ "schema": "com.plum/ncx10-health-check/1",
  "data": {
    "factory": "Factory A",
    "serialNumber": "EU3571",
    "status": "RUNNING",
    "lastStartedAt": 1539598697944,
    "temperature": 34.56,
    "endOfLife": false,
    "floorNumber": { "int": 2 }
  }
}
```

This is better: it's significantly less compact than the binary representation, but it is human-readable from top to bottom. The only oddity is that the overall payload is in fact no longer an Avro: it *is* valid JSON, and the data portion is a valid JSON-format Avro, but the overall payload is *not* an Avro. This is because there is no Avro schema that models the entire payload; instead to process an instance of this event, we would do the following:

1 Parse the event initially as JSON to extract the schema string.
2 Retrieve the Avro schema per the schema string.
3 Use the schema string to retrieve the JSON node for the event's data.
4 Deserialize the JSON data node into a Check POJO object by using the retrieved Avro schema.

Don't worry if the nuances of self-describing Avro seem abstract right now. We will put these ideas into practice in the next chapter; things should soon seem more concrete.

Before we move on, let's find a home for all of our schemas: a schema registry.

### 6.3.3 Plum's schema registry

In the preceding example, we defined a health-check event for the NCX-10 machine by using the Avro schema language, and then embedded this Avro definition inside our Java application. If we took this to its logical conclusion, we would be embedding this Avro definition file into every app that needed to understand NCX-10 health-check events. Figure 6.7 illustrates this process.

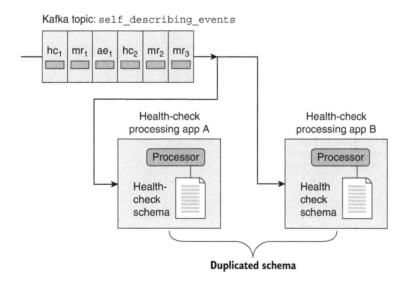

Figure 6.7 We have two Kafka applications that want to process health-check events. Both of these apps contain a definition of the health-check event in Avro; there is no central source of truth for this schema.

This copy-and-paste approach is a bad idea, because we don't have a *single source of truth* for the definition of an NCX-10 health-check event within Plum. Instead, we have multiple definitions of this event, scattered around multiple codebases. We have to hope that all the definitions are the same, or Plum will experience the following:

- Self-describing events being written to Kafka with contradictory data structures for the same schema
- Runtime failures in our Kafka consuming applications as they expect one event structure and get another

Plum needs a single source of truth for all of our schemas—a single location that registers our schemas, and that all teams can access to understand the definition of each schema. Figure 6.8 visualizes the schema registry for Plum, containing three event schemas.

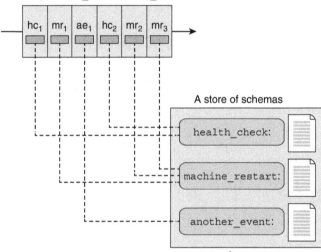

**Figure 6.8   Plum now has a schema registry that contains the master copy of the schema for each type of event. All consuming and producing apps should use these master copies.**

At its simplest, a schema registry can be just a shared folder, perhaps on S3, HDFS, or NFS. The schema syntax we defined earlier maps nicely onto a folder structure, here in Amazon S3:

```
s3://plum-schemas/com.plum/ncx10-health-check/1
```

The file at the preceding path would be the Avro definition file for version 1 of the NCX-10 health-check event.

Going beyond a simple shared folder structure, there are two actively developed open source schema registries:

- *Confluent Schema Registry* (https://github.com/confluentinc/schema-registry)— An integral part of the Confluent Platform for Kafka-based data pipelines.

Confluent Schema Registry supports Avro only, with first-class support for Avro's schema evolution. It uses Kafka as the underlying storage mechanism and is a distributed system with a single master architecture. It assigns a registry-unique ID (monotonically increasing) to each registered schema.

- *Iglu* (https://github.com/snowplow/iglu)—An integral part of the Snowplow open source event data pipeline. Iglu supports multiple schema technologies, including Avro, JSON Schema, and Thrift. It supports schema resolution across multiple schema registries and uses semantic URIs to address schemas. Iglu is used in Snowplow but intended to be general-purpose (with Scala and Objective-C client libraries).

Putting either of these schema registries into service for Plum is beyond the scope of this book, but we encourage you to check them out.

## Summary

- A unified log is a *decoupled* architecture: consumers and producers of event streams have no particular knowledge of each other.
- The contract between event consumers and producers is represented by the schema of each event.
- Schema technologies we can use include JSON Schema, Apache Avro, Thrift, and protocol buffers. Avro is a good choice.
- We can define our event schemas by using Avro's JSON-based schema syntax, and then use code-generation to build Java bindings (think POJOs) to the schemas.
- Avro has a JSON representation for data, and a binary representation. We wrote a simple Java app to convert the JSON representation into Java and back into binary.
- We need a way to associate events with their schemas: either one stream per schema, trial and error, or self-describing events.
- We can model a simple self-describing event for Avro by using Avro itself. We need slightly different implementations for the binary- and JSON-based representations.
- A schema registry is a central repository for all of our schemas: this is the source of truth for which schemas are available as part of our company's unified log.

# Archiving events 7

**This chapter covers**

- Why you should be archiving raw events from your unified log
- The what, where, and how of archiving
- Archiving events from Kafka to Amazon S3
- Batch-processing an event archive via Spark and Elastic MapReduce

So far, our focus has been on reacting to our events *in stream*, as these events flow through our unified log. We have seen some great near-real-time use cases for these event streams, including detecting abandoned shopping carts and monitoring our servers. You would be correct in thinking that the immediacy of a unified log is one of its most powerful features.

But in this chapter, we will take a slight detour and explore another path that our event streams can take: into a long-term store, or *archive*, of all our events. To continue with the flowing water analogies so beloved of data engineers: if the unified log is our Mississippi River, our event archive is our bayou:[1] a sleepy but vast backwater, fertile for exploration.

---

[1] A more formal definition of bayou can be found at Wikipedia: https://en.wikipedia.org/wiki/Bayou.

There are many good reasons to archive our events like this; we will make the case for these first in an *archivist's manifesto*. With the case made, we will then introduce the key building blocks of a good event archive: which events to archive, where to store them, and what tooling to use to achieve this.

There is no substitute for implementing an actual archive, so we will follow up the theory with some archiving of the shopping events on the Nile website, as introduced in chapter 2. For this, we will use Secor, a tool from Pinterest that can mirror Kafka topics (such as our Nile raw event stream) to a bucket in Amazon S3.

After we have our shopping events safely stored in Amazon S3, we can start to mine that archive for insights. We can use a small set of open source batch-processing frameworks to do this—the most well-known of these are Apache Hadoop and Apache Spark. We will write a simple analytics job on Nile's shopping events by using Spark, coding our job interactively using the Spark console (a nice feature of Spark). Finally, we will then "operationalize" our Spark job by running it on a distributed server cluster by using Amazon's Elastic MapReduce service.

So, plenty to learn and do! Let's get started.

## 7.1 The archivist's manifesto

So far, we have always worked directly on the event streams flowing through our unified log: we have created apps that wrote events to Apache Kafka or Amazon Kinesis, read events from Kafka or Kinesis, or did both. The unified log has proved a great fit for the various near-real-time use cases we have explored so far, including enriching events, detecting abandoned shopping carts, aggregating metrics, and monitoring systems.

But there is a limit to this approach: neither Kafka nor Kinesis is intended to contain your entire event archive. In Kinesis, the *trim horizon* is set to 24 hours and can be increased up to 1 week (or 168 hours). After that cutoff, older events are *trimmed*—deleted from the stream forever. With Apache Kafka, the trim horizon (called the *retention period*) is also configurable: in theory, you could keep all your event data inside Kafka, but in practice, most people would limit their storage to one week (which is the default) or a month. Figure 7.1 depicts this limitation of our unified log.

As unified log programmers, the temptation is to say, "If data will be trimmed after some hours or days, let's just make sure we do all of our processing before that time window is up!" For example, if we need to calculate certain metrics in that time window, let's ensure that those metrics are calculated and safely written to permanent storage in good time. Unfortunately, this approach has key shortcomings; we could call these the *Three Rs*:

- Resilience
- Reprocessing
- Refinement

Let's take these in turn.

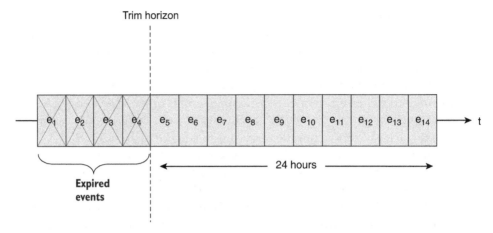

**Figure 7.1   The Kinesis trim horizon means that events in our stream are available for processing for only 24 hours after first being written to the stream.**

### 7.1.1  *Resilience*

We want our unified log processing to be as resilient in the face of failure as possible. If we have Kinesis or Kafka set to delete events forever after 24 hours, that makes our event pipeline much more fragile: we must fix any processing failures before those events are gone from our unified log forever. If the problem happens over the weekend and we do not detect it, we are in trouble; there will be permanent gaps in our data that we have to explain to our boss, as visualized in figure 7.2.

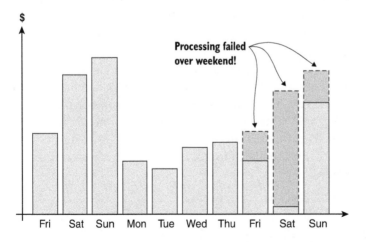

**Figure 7.2   A dashboard of daily sales annotated to explain the data missing from the second weekend. From the missing data, we can surmise that the outage started toward the end of Friday, continued through the weekend, and was identified and fixed on Monday (allowing most of Sunday's data to be recovered).**

Remember too that our pipelines typically consist of multiple processing stages, so any stages downstream of our failure will also be missing important input data. We'll have a *cascade failure*. Figure 7.3 demonstrates this cascade failure—our upstream job that validates and enriches our events fails, causing cascade failures in various downstream applications, specifically these:

- A stream processing job that loads the events into Amazon Redshift
- A stream processing job that provides management dashboards, perhaps similar to that visualized in figure 7.2
- A stream processing job that monitors the event stream looking for customer fraud

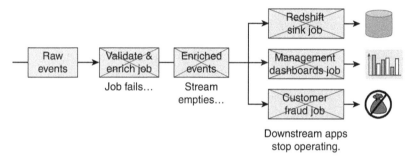

**Figure 7.3   A failure in our upstream job that validates and enriches our event causes cascade failures in all of our downstream jobs, because the stream of enriched events that they depend on is no longer being populated.**

Even worse, with Kinesis, the window we have to recover from failure is shorter than 24 hours. This is because, after the problem is fixed, we will have to resume processing events from the point of failure onward. Because we can read only up to 2 MB of data from each Kinesis shard per second, it may not be physically possible to process the entire backlog before some of it is trimmed (lost forever). In those cases, you might want to increase the trim horizon to the maximum allowed, which is 168 hours.

Because we cannot foresee and fix all the various things that could fail in our job, it becomes important that we have a robust backup of our incoming events. We can then use this backup to recover from any kind of stream processing failure at our own speed.

### 7.1.2   Reprocessing

In chapter 5, we wrote a stream processing job in Samza that detected abandoned shopping carts after 30 minutes of customer inactivity. This worked well, but what if we have a nagging feeling that a different definition of cart abandonment might suit our business better? If we had all of our events stored somewhere safe, we could apply multiple different cart abandonment algorithms to that event archive, review the results, and then port the most promising algorithms into our stream processing job.

More broadly, there are several reasons that we might want to reprocess an event stream from a full archive of that stream:

- We want to fix a bug in an existing calculation or aggregation. For example, we find that our daily metrics are being calculated against the wrong time zone.
- We want to alter our definition of a metric. For example, we decide that a user's browsing session on our website ends after 15 minutes of inactivity, not 30 minutes.
- We want to apply a new calculation or aggregation retrospectively. For example, we want to track cumulative e-commerce sales per device type as well as per country.

All of these use cases depend on having access to the event stream's entire history—which, as you've seen, isn't possible with Kinesis, and is somewhat impractical with Kafka.

### 7.1.3  *Refinement*

Assuming that the calculations and aggregations in our stream processing job are bug-free, just how accurate will our job's results be? They may not be as accurate as we would like, for three key reasons:

- *Late-arriving data*—At the time that our job is performing the calculation, it may not have access to all of the data that the calculation needs.
- *Approximations*—We may choose to use an approximate calculation in our stream processing job on performance grounds.
- *Framework limitations*—Architectural limitations may be built into our stream processing framework that impact the accuracy of our processing.

#### LATE-ARRIVING DATA

In the real world, data is often *late arriving*, and this has an impact on the accuracy of our stream processing calculations. Think of a mobile game that is sending to our unified log a stream of events recording each player's interaction with the game.

Say a player takes the city's underground metro and loses signal for an hour. The player continues playing on the metro, and the game keeps faithfully recording events; the game then sends them the cached events in one big batch when the user gets above ground again. Unfortunately, while the user was underground, our stream processing job already decided that the player had finished playing the game and updated its metrics on completed game sessions accordingly. The late-arriving batch of events invalidates the conclusions drawn by our stream processing app, as visualized in figure 7.4.

The data that online advertising companies provide to their customers is another example. It takes days (sometimes weeks) for these companies to determine which clicks on ads were real and which ones were fraudulent. Therefore, it also takes days or weeks for marketing spend data to be finalized, sometimes referred to as *becoming golden*. As a result, any join we do in a stream processing job between our ad click and

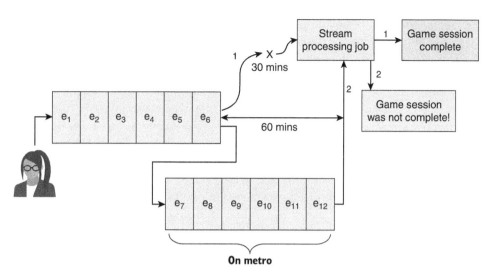

**Figure 7.4  Our stream processing job draws the wrong conclusion—that a game session has finished— because relevant events arrive too late to be included in the decision-making process.**

what we paid for that click will be only a first estimate, and subject to refinement using our late-arriving data.

### APPROXIMATIONS

For performance reasons, you may choose to make approximations in your stream processing algorithms. A good example, explored in *Big Data*, is calculating unique visitors to a website. Calculating unique visitors (COUNT DISTINCT in SQL) can be challenging because the metric is not additive—for example, you cannot calculate the number of unique visitors to a website in a month by adding together the unique visitor numbers for the constituent weeks. Because accurate uniqueness counts are computationally expensive, often we will choose an approximate algorithm such as HyperLogLog for our stream processing.[2]

Although these approximations are often satisfactory for stream processing, an analytics team usually will want to have the option of refining those calculations. In the case of unique visitors to the website, the analytics team might want to be able to generate a true COUNT DISTINCT across the full history of website events.

### FRAMEWORK LIMITATIONS

This last reason is a hot topic in stream processing circles. Nathan Marz's Lambda Architecture is designed around the idea that the stream processing component (what he calls the *speed layer*) is inherently unreliable.

A good example of this is the concept of *exactly-once* versus *at-least-once* processing, which has been discussed in chapter 5. Currently, Amazon Kinesis offers only

---

[2]  The HyperLogLog algorithm for approximating uniqueness counts is described at Wikipedia: https://en .wikipedia.org/wiki/HyperLogLog.

at-least-once processing: an event will never be lost as it moves through a unified log pipeline, but it may be duplicated one or more times. If a unified log pipeline duplicates events, this is a strong rationale for performing additional refinement, potentially using a technology that supports exactly-once processing. It is worth noting, though, that as of release 0.11, Apache Kafka also supports exactly-once delivery semantics.

This is a hot topic because Jay Kreps, the original architect of Apache Kafka, disagrees that stream processing is inherently unreliable; he sees this more as a transitional issue with the current crop of unified log technologies.[3] He has called this the *Kappa Architecture*, the idea that we should be able to use a single stream-processing stack to meet all of our unified log needs.

My view is that framework limitations will ultimately disappear, as Jay says. But late-arriving data and pragmatic approximation decisions will always be with us. And today these three accuracy issues collectively give us a strong case for creating and maintaining an event archive.

## 7.2    A design for archiving

In the preceding section, you saw that archiving the event streams that flow through our unified log makes our processing architecture more robust, allows us to reprocess when needed, and lets us refine our processing outputs. In this section, we will look at the what, where, and how of event stream archiving.

### 7.2.1    What to archive

We have decided that archiving events within our unified log is a good idea, but what precisely should we be archiving? The answer may not be what you are expecting: you should archive the *rawest events* you can, as *far upstream* in your event pipeline as possible. This early archiving is shown in figure 7.5, which builds on the unified log topology set out earlier in figure 7.3.

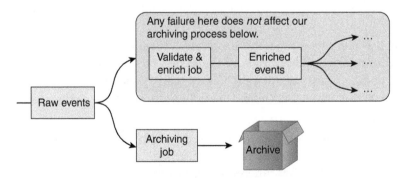

**Figure 7.5   By archiving as far upstream as possible, we insulate our archiving process from any failures that occur downstream of the raw event stream.**

---

[3]  You can read more about Kreps' ideas in his seminal article that coined the term "Kappa Architecture" at www.oreilly.com/ideas/questioning-the-lambda-architecture.

To give two examples from earlier chapters:

- In chapters 2 and 3, we would archive the three types of events generated by shoppers on the Nile website.
- In chapter 5, we would archive the health-check event generated by the machines on Plum's production line.

Event validation and enrichment can be a costly process, so why insist on archiving the rawest events? Again, it comes down to the three Rs:

- *Resilience*—By archiving as upstream as possible, we are guaranteeing that there are no intermediate stream-processing jobs that could break and thus cause our archiving to fail.
- *Reprocessing*—We may want to reprocess *any* part of our event pipeline—yes, even the initial validation and enrichment jobs. By having the rawest events archived, we should be able to reprocess anything downstream.
- *Refinement*—Any refinement process (such as a Hadoop or Spark batch-processing job) should start from the exact same input events as the stream processing job that it aims to refine.

### 7.2.2 *Where to archive*

We need to archive our event stream to permanent file storage that has the following characteristics:

- Is robust, because we don't want to learn later that parts of the archive have been lost
- Makes it easy for data processing frameworks such as Hadoop or Spark to quickly load the archived events for further processing or refinement

Both requirements point strongly to a distributed filesystem. Table 7.1 lists the most popular examples; our rawest event stream should be archived to at least one of these.

**Table 7.1  Examples of distributed filesystems**

| Distributed filesystem | Hosted? | API | Description |
| --- | --- | --- | --- |
| Amazon Simple Storage Service (S3) | Yes | HTTP | A hosted file storage service, part of Amazon Web Services. |
| Azure Blob Storage | Yes | HTTP | A hosted unstructured data storage service, part of Microsoft Azure. |
| Google Cloud Storage | Yes | HTTP | A hosted object storage service, part of Google Cloud Platform. |
| Hadoop Distributed File System (HDFS) | No | Java, Thrift | A distributed filesystem written in Java for the Hadoop framework. |
| OpenStack Swift | No | HTTP | A distributed, highly available, eventually consistent object store. |

**Table 7.1   Examples of distributed filesystems *(continued)***

| Distributed filesystem | Hosted? | API | Description |
|---|---|---|---|
| Riak Cloud Storage (CS) | No | HTTP | Built on the Riak database. API compatible with Amazon S3. |
| Tachyon | No | Java, Thrift | A memory-centric storage system optimized for Spark and Hadoop processing. Implements the HDFS interface. |

Your choice of storage will be informed by whether your company uses an infrastructure-as-a-service (IaaS) offering such as AWS, Azure, or Google Compute Engine, or has built its own data processing infrastructure. But even if your company has built its own private infrastructure, you may well choose to archive your unified log to a hosted service such as Amazon S3 as well, as an effective off-site backup.

### 7.2.3   *How to archive*

The fundamentals of archiving events from our unified log into our permanent file storage are straightforward. We need a stream consumer that does the following:

- *Reads* from each shard or topic in our stream
- *Batches* a sensible number of events into a file that is optimized for subsequent data processing
- *Writes* each file to our distributed filesystem of choice

The good news is that this is a solved problem, so we won't have to write any code just yet! Various companies have open sourced tools to archive either Kafka or Kinesis event streams to one or other of these distributed filesystems. Table 7.2 itemizes the most well-known of these.

**Table 7.2   Tools for archiving our unified log from a distributed filesystem**

| Tool | From | To | Creator | Description |
|---|---|---|---|---|
| Camus | Kafka | HDFS | LinkedIn | A MapReduce job that loads Kafka into HDFS. Can autodiscover Kafka topics. |
| Flafka | Kafka | HDFS | Cloudera / Flume | Part of the Flume project. Involves configuring a Kafka source plus HDFS sink. |
| Bifrost | Kafka | S3 | uSwitch.com | Writes events to S3 in uSwitch.com's own baldr binary file format. |
| Secor | Kafka | S3 | Pinterest | A service for persisting Kakfa topics to S3 as Hadoop SequenceFiles. |
| kinesis-s3 | Kinesis | S3 | Snowplow | A Kinesis Client Library application to write a Kinesis stream to S3. |
| Connect S3 | Kafka | S3 | Confluent | Allows exporting data from Kafka topics to S3 objects in either Avro or JSON formats. |

Now that you understand the why and how of archiving our raw event stream, we will put the theory into practice by using one of these tools: Pinterest's Secor.

## 7.3    Archiving Kafka with Secor

Let's return to Nile, our online retailer from chapter 5. Remember that Nile's shoppers generate three types of events, all of which are collected by Nile and written to their Apache Kafka unified log. The three event types are as follows:

- Shopper views product
- Shopper adds item to cart
- Shopper places order

Alongside the existing stream processing we are doing on these events, Nile wants to archive all of these events to Amazon S3. Figure 7.6 sets out the desired end-to-end architecture.

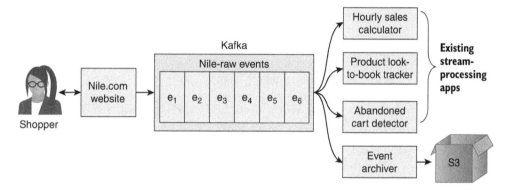

**Figure 7.6    Alongside Nile's three existing stream-processing applications, we will be adding a fourth application, which archives the raw event stream to Amazon S3.**

Looking at the various archiving tools in table 7.2, the only three that fit the bill for archiving Kafka to Amazon S3 are uSwitch's Bifrost, Pinterest's Secor, and Confluent's Kafka Connect S3. For Nile, we choose Secor over Bifrost because Secor's storage format—Hadoop SequenceFiles—is much more widely adopted than Bifrost's baldr format. Remember, Nile will want to mine the event data stored in its S3 archive for many years to come, so it's crucial that we use non-niche data formats that will be well supported for the foreseeable future. We also could have chosen Confluent's Kafka Connect S3, but that would require us to install Confluent's Kafka distribution instead of Apache's. Because we already have the latter running, let's get started with that.

### 7.3.1  *Warming up Kafka*

First, we need to get Nile's raw events flowing into Kafka again. Assuming you still have Kafka deployed as per chapter 2, you can start up Apache ZooKeeper again like so:

```
$ cd kafka_2.12-2.0.0
$ bin/zookeeper-server-start.sh config/zookeeper.properties
```

Now we are ready to start Kafka in a second terminal:

```
$ cd kafka_2.12-2.0.0
$ bin/kafka-server-start.sh config/server.properties
```

Let's create the `raw-events-ch07` topic in Kafka as we did in chapter 5, so we can send some events in straightaway. You can make that topic available in Kafka by creating it like this:

```
$ bin/kafka-topics.sh --create --topic raw-events-ch07 \
  --zookeeper localhost:2181 --replication-factor 1 --partitions 1
Created topic "raw-events-ch07".
```

Now in your third terminal, run the Kafka console producer like so:

```
$ bin/kafka-console-producer.sh \
    --broker-list localhost:9092 --topic raw-events-ch07
```

This producer will sit waiting for input. Let's feed it some events, making sure to press Enter after every line:

```
{ "subject": { "shopper": "789" }, "verb": "add", "indirectObject":
 "cart", "directObject": { "item": { "product": "aabattery", "quantity":
 12, "price": 1.99 }}, "context": { "timestamp": "2018-10-30T23:01:29" } }

{ "subject": { "shopper": "456" }, "verb": "add", "indirectObject":
 "cart", "directObject": { "item": { "product": "thinkpad", "quantity":
 1, "price": 1099.99 }},"context": { "timestamp": "2018-10-30T23:03:33" }}

{ "subject": { "shopper": "789" }, "verb": "add", "indirectObject":
 "cart", "directObject": { "item": { "product": "ipad", "quantity": 1,
 "price": 499.99 } }, "context": { "timestamp": "2018-10-30T00:04:41" } }

{ "subject": { "shopper": "789" }, "verb": "place", "directObject":
 { "order": { "id": "123", "value": 511.93, "items": [ { "product":
 "aabattery", "quantity": 6 }, { "product": "ipad", "quantity": 1} ] } },
 "context": { "timestamp": "2018-10-30T00:08:19" } }

{ "subject": { "shopper": "123" }, "verb": "add", "indirectObject":
 "cart", "directObject": { "item": { "product": "skyfall", "quantity":
 1, "price": 19.99 }}, "context": { "timestamp": "2018-10-30T00:12:31" } }

{ "subject": { "shopper": "123" }, "verb": "add", "indirectObject":
 "cart", "directObject": { "item": { "product": "champagne", "quantity":
 5, "price": 59.99 }}, "context": { "timestamp": "2018-10-30T00:14:02" } }
```

```
{ "subject": { "shopper": "123" }, "verb": "place", "directObject":
{ "order": { "id": "123", "value": 179.97, "items": [ { "product":
"champagne", "quantity": 3 } ] } }, "context": { "timestamp":
"2018-10-30T00:17:18" } }
```

Phew! After entering all of these, you can now press Ctrl-D to exit the console producer. It's a little hard to tell from all those JSON objects exactly what is happening on the Nile website. Figure 7.7 illustrates the three shoppers who are adding these products to their carts and then placing their orders.

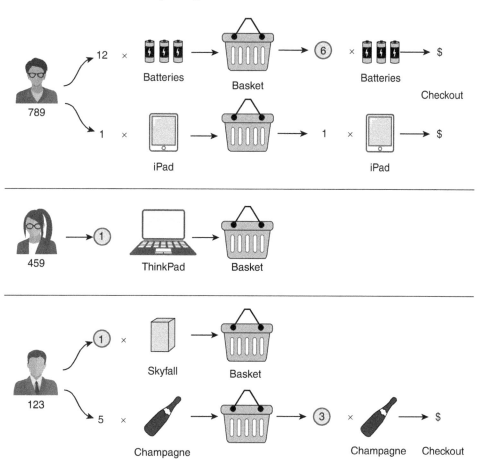

**Figure 7.7  Our seven events visualized: three shoppers are adding products to their basket; two of those shoppers are going on to place orders, but with reduced quantities in their baskets.**

In any case, those seven events should all be safely stored in Kafka now, and we can check this easily using the console consumer:

```
$ bin/kafka-console-consumer.sh --topic raw-events-ch07 --from-beginning \
    --bootstrap-server localhost:9092
```

```
{ "subject": { "shopper": "789" }, "verb": "add", "indirectObject":
"cart", "directObject": { "item": { "product": "aabattery", "quantity":
12, "unitPrice": 1.99 } }, "context": { "timestamp":
"2018-10-30T23:01:2" } }
...
```

Then press Ctrl-C to exit the consumer. Great—our events have been safely logged in Kafka, so we can move on to the archiving.

### 7.3.2 *Creating our event archive*

Remember that Nile wants all of the raw events that are being written to Kafka to be archived to Amazon Simple Storage Service, more commonly referred to as Amazon S3. From earlier in part 2, you should be comfortable with Amazon Web Services, although we have yet to use Amazon S3.

In chapter 4, we created an Amazon Web Services user called `ulp`, and gave that user full permissions on Amazon Kinesis. We now need to log back into the AWS as our root user and assign the `ulp` user full permissions on Amazon S3. From the AWS dashboard:

1  Click the Identity & Access Management icon.
2  Click the Users option in the left-hand navigation pane.
3  Click your `ulp` user.
4  Click the Add Permissions button.
5  Click the Attach Existing Policies Directly tab.
6  Select the AmazonS3FullAccess policy and click Next: Review.
7  Click the Add Permissions button.

We are now ready to create an Amazon S3 *bucket* to archive our events into. An S3 bucket is a top-level folder-like resource, into which we can place individual files. Slightly confusingly, the names of S3 buckets have to be globally unique. To prevent your bucket's name from clashing with that of other readers of this book, let's adopt a naming convention like this:

```
s3://ulp-ch07-archive-{{your-first-pets-name}}
```

Use the AWS CLI tool's `s3` command and `mb` (for *make bucket*) subcommand to create your new bucket, like so:

```
$ aws s3 mb s3://ulp-ch07-archive-little-torty --profile=ulp
make_bucket: s3://ulp-ch07-archive-little-torty/
```

Our bucket has been created. We now have our shopper events sitting in Kafka and an empty bucket in Amazon S3 ready to archive our events in. Let's add in Secor and connect the dots.

### 7.3.3   *Setting up Secor*

There are no prebuilt binaries for Secor, so we will have to build it from source our-selves. The Vagrant development environment has all the tools we need to get started:

```
$ cd /vagrant
$ wget https://github.com/pinterest/secor/archive/v0.26.tar.gz
$ cd secor-0.26
```

Next, we need to edit the configuration files that Secor will run against. First, load this file in your editor of choice:

```
/vagrant/secor/src/main/config/secor.common.properties
```

Now update the settings within the MUST SET section, as in the following listing.

---

**Listing 7.1   secor.common.properties**

```
...
# Regular expression matching names of consumed topics.          Restricts our archiving
secor.kafka.topic_filter=raw-events-ch07                         to only Nile's raw events
                                                              ◄── from this chapter

# AWS authentication credentials.          Same as aws_access_key_id
aws.access.key={{access-key}}          ◄── in ~/.aws/credentials
aws.secret.key={{secret-key}}          ◄─┐ Same as aws_secret_access_key
...                                      └ in ~/.aws/credentials
```

---

Next you need to edit this file:

```
/vagrant/secor/src/main/config/secor.dev.properties
```

We have only one setting to change here: the secor.s3.bucket property. This needs to match the bucket that we set up in section 7.3.2. When that's done, your secor.dev.properties file should look similar to that set out in the following listing.

---

**Listing 7.2   secor.dev.properties**

```
include=secor.common.properties          ◄─┐ Imports the file we
                                           └ edited previously

############
# MUST SET #
############

# Name of the s3 bucket where log files are stored.
secor.s3.bucket=ulp-ch07-archive-{{your-first-pets-name}}     ◄─┐ Set to your
                                                                └ bucket's name
################
# END MUST SET #
################

kafka.seed.broker.host=localhost
kafka.seed.broker.port=9092
```

```
zookeeper.quorum=localhost:2181

# Upload policies.        ⊲──┤  Rules for uploading
# 10K                        │  files to S3
secor.max.file.size.bytes=10000
# 1 minute
secor.max.file.age.seconds=60
```

From this listing, we can see that the default rules for uploading our event files to S3 are to wait for either 1 minute or until our file contains 10,000 bytes, whichever comes sooner. These defaults are fine, so we will leave them as is. And that's it; we can leave the other configuration files untouched and move on to building Secor:

```
$ mvn package
...
[INFO] BUILD SUCCESS...
...
$ sudo mkdir /opt/secor
$ sudo tar -zxvf target/secor-0.26-SNAPSHOT-bin.tar.gz -C /opt/secor
...
lib/jackson-core-2.6.0.jar
lib/java-statsd-client-3.0.2.jar
```

Finally, we are ready to run Secor:

```
$ sudo mkdir -p /mnt/secor_data/logs
$ cd /opt/secor
$ sudo java -ea -Dsecor_group=secor_backup \
    -Dlog4j.configuration=log4j.prod.properties \
    -Dconfig=secor.dev.backup.properties -cp \
    secor-0.26.jar:lib/* com.pinterest.secor.main.ConsumerMain
Nov 05, 2018 11:26:32 PM com.twitter.logging.Logger log
INFO: Starting LatchedStatsListener
...
INFO: Cleaning up!
```

Note that some few seconds will elapse before the final INFO: Cleaning up! message appears; in this time, Secor is finalizing the batch of events, storing it in a Hadoop SequenceFile, and uploading it to Amazon S3.

Let's quickly check that the file has successfully uploaded to S3. From the AWS dashboard:

1 Click the S3 icon.
2 Click the bucket ulp-ch07-archive-{{your-first-pets-name}}.
3 Click each subfolder until you arrive at a single file.

This file contains our seven events, read from Kafka and successfully uploaded to S3 by Secor, as you can see in figure 7.8.

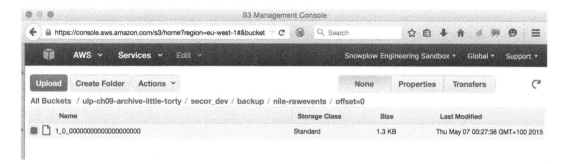

**Figure 7.8   The AWS UI for Amazon S3, showing our archived events**

We can download the archived file by using the AWS CLI tools:

```
$ cd /tmp
$ PET=little-torty
$ FILE=secor_dev/backup/raw-events-ch07/offset=0/1_0_00000000000000000000
$ aws s3 cp s3://ulp-ch07-archive-${PET}/${FILE} . --profile=ulp
download: s3://ulp-ch07-archive-little-torty/secor_dev/backup/raw-events-
ch07/offset=0/1_0_00000000000000000000 to ./1_0_00000000000000000000
```

What's in the file? Let's have a quick look inside it:

```
$ file 1_0_00000000000000000000
1_0_00000000000000000000: Apache Hadoop Sequence file version 6
$ $ head -1 1_0_00000000000000000000
SEQ!org.apache.hadoop.io.LongWritable"org.a pache.hadoop.io.BytesWritabl...
```

Alas, we can't easily read the file from the command line. The file is stored by Secor as a Hadoop SequenceFile, which is a flat file format consisting of binary key-value pairs.[4] Batch processing frameworks such as Hadoop can easily read SequenceFiles, and that is what we'll explore next.

## 7.4    Batch processing our archive

Now that we have our raw events safely archived in Amazon S3, we can use a *batch processing framework* to process these events in any way that makes sense to Nile.

### 7.4.1    Batch processing 101

The fundamental difference between batch processing frameworks and stream processing frameworks relates to the way in which they ingest data. Batch processing frameworks expect to be run against a *terminated set of records*, unlike the unbounded event stream (or streams) that a stream-processing framework reads. Figure 7.9 illustrates a batch processing framework.

---

[4]   A definition of the Hadoop SequenceFile format is available at https://wiki.apache.org/hadoop/Sequence-File.

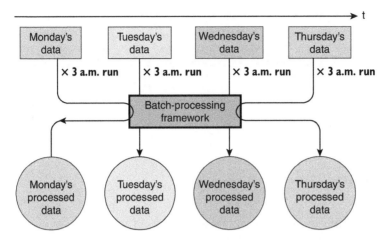

**Figure 7.9   A batch processing framework has processed four distinct batches of events, each belonging to a different day of the week. The batch processing framework runs at 3 a.m. daily, ingests the data for the prior day from storage, and writes its outputs back to storage at the end of its run.**

By way of comparison, figure 7.10 illustrates how a stream processing framework works on an unterminated stream of events.

A second, perhaps more historical, difference is that batch processing frameworks have been used with a much broader variety of data than stream processing frameworks. The canonical example for batch processing as popularized by Hadoop is counting words in a corpus of English-language documents (semistructured data). By contrast, stream processing frameworks have been more focused on well-structured event stream data, although some promising initiatives support processing other data types in stream.[5]

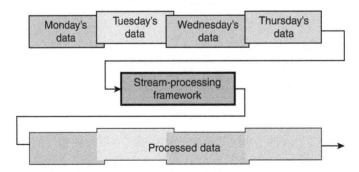

**Figure 7.10   A stream processing framework doesn't distinguish any breaks in the incoming event stream. Monday through Thursday's data exists as one unbounded stream, likely with overlap due to late-arriving events.**

---

[5]  An experimental initiative to integrate Luwak within Samza can be found at https://github.com/romseygeek/samza-luwak.

Table 7.3 lists the major distributed batch-processing frameworks. Of these, Apache Hadoop and, increasingly, Apache Spark are far more widely used than Disco or Apache Flink.

**Table 7.3  Examples of distributed batch-processing frameworks**

| Framework | Started | Creator | Description |
|---|---|---|---|
| Disco | 2008 | Nokia Research Center | MapReduce framework written in Erlang. Has its own filesystem, DDFS. |
| Apache Flink | 2009 | TU Berlin | Formerly known as Project Stratosphere, Flink is a streaming dataflow engine with a DataSet API for batch processing. Write jobs in Scala, Java, or Python. |
| Apache Hadoop | 2008 | Yahoo! | Software framework written in Java for distributed processing (Hadoop MapReduce) and distributed storage (HDFS). |
| Apache Spark | 2009 | UC Berkeley AMPLab | Large-scale data processing, supporting cyclic data flow and optimized for memory use. Write jobs in Scala, Java, or Python. |

What do we mean when we say that these frameworks are *distributed*? Simply put, we mean that they have a master-slave architecture:

- The master supervises the slaves and parcels out units of work to the slaves.
- The slaves (sometimes called *workers*) receive the units of work, perform them, and provide status updates to the master.

Figure 7.11 represents this architecture. This distribution allows processing to scale horizontally, by adding more slaves.

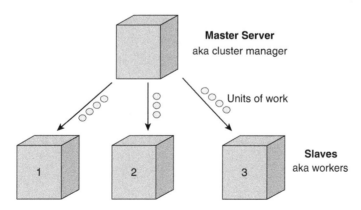

**Figure 7.11  In a distributed data-processing architecture, the master supervises a set of slaves, allocating them units of work from the batch processing job.**

### 7.4.2    *Designing our batch processing job*

The analytics team at Nile wants a report on the lifetime behavior of every Nile shopper:

- How many items has each shopper added to their basket, and what is the total value of all items added to basket?
- Similarly, how many orders has each shopper placed, and what is the total value of each shopper's orders?

Figure 7.12 shows a sample report.

| Shopper | Added to basket | | Placed order | |
|---|---|---|---|---|
| | Items | Value | Orders | Value |
| 123 | 17 | $ 107.98 | 1 | $ 37.99 |
| 456 | 1 | $ 499.99 | 1 | $ 499.99 |
| 789 | 4 | $ 368.04 | 0 | $ 0 |

Figure 7.12  For each shopper, the Nile analytics team wants to know the volume and value of items added to basket, and the volume and value of orders placed. With this data, the value of abandoned carts is easily calculated.

If we imagine that the Nile analytics team came up with this report six months into us operating our event archive, we can see that this is a classic piece of *reprocessing*: Nile wants us to apply new aggregations retrospectively to the full event history.

Before we write a line of code, let's come up with the algorithms required to produce this report. We will use SQL-esque syntax to describe the algorithms. Here's an algorithm for the shoppers' add-to-basket activity:

```
GROUP BY shopper_id
WHERE event_type IS add_to_basket
  items = SUM(item.quantity)
  value = SUM(item.quantity * item.price)
```

We calculate the volume and value of items added to each shopper's basket by looking at the *Shopper adds item to basket* events. For the volume of items, we sum all of the item quantities recorded in those events. Value is a little more complex: we have to multiply all of the item quantities by the item's unit price to get the total value.

The shoppers' order activity is even simpler:

```
GROUP BY shopper_id
WHERE event_type IS place_order
  orders = COUNT(rows)
  value  = SUM(order.value)
```

For each shopper, we look at their *Shopper places order* events only. A simple count of those rows tells us the number of orders they have placed. A sum of the values of those orders gives us the total amount they have spent.

And that's it. Now that you know what calculations we want to perform, we are ready to pick a batch processing framework and write them.

### 7.4.3 *Writing our job in Apache Spark*

We need to choose a batch processing framework to write our job in. We will use Apache Spark: it has an elegant Scala API for writing the kinds of aggregations that we require, and it plays relatively well with Amazon S3, where our event archive lives, and Amazon Elastic MapReduce (EMR), which is where we will ultimately run our job. Another plus for Spark is that it's easy to build up our processing job's logic interactively by using the Scala console.

Let's get started. We are going to create our Scala application by using Gradle. First, create a directory called spark, and then switch to that directory and run the following:

```
$ gradle init --type scala-library
...
BUILD SUCCESSFUL
...
```

As we did in previous chapters, we'll now delete the stubbed Scala files that Gradle created:

```
$ rm -rf src/*/scala/*
```

The default build.gradle file in the project root isn't quite what we need either, so replace it with the code in the following listing.

**Listing 7.3  build.gradle**

```
apply plugin: 'scala'

configurations {
    provided                              ◁──┐  This lets us specify
}                                             │  dependencies as provided,
                                              │  so they won't be included
sourceSets {                          ◁───────┘  in an assembled JAR.
    main.compileClasspath += configurations.provided
}

repositories {
  mavenCentral()
}

version = '0.1.0'

  ScalaCompileOptions.metaClass.daemonServer = true
  ScalaCompileOptions.metaClass.fork = true
  ScalaCompileOptions.metaClass.useAnt = false
  ScalaCompileOptions.metaClass.useCompileDaemon = false
```

```
dependencies {
  runtime "org.scala-lang:scala-compiler:2.12.7"
  runtime "org.apache.spark:spark-core_2.12:2.4.0"
  runtime "org.apache.spark:spark-sql_2.12:2.4.0"
  compile "org.scala-lang:scala-library:2.12.7"
  provided "org.apache.spark:spark-core_2.12:2.4.0"
  provided "org.apache.spark:spark-sql_2.12:2.4.0"
}

jar {
  dependsOn configurations.runtime
  from {
    (configurations.runtime - configurations.provided).collect {
      it.isDirectory() ? it : zipTree(it)
    }
  } {
    exclude "META-INF/*.SF"
    exclude "META-INF/*.DSA"
    exclude "META-INF/*.RSA"
  }
}

task repl(type:JavaExec) {
    main = "scala.tools.nsc.MainGenericRunner"
    classpath = sourceSets.main.runtimeClasspath
    standardInput System.in
    args '-usejavacp'
}
```

**We need Spark and Spark SQL as dependencies.**

**Make sure to exclude the "provided" dependencies when constructing our fat jar.**

**Task to start the Scala console**

With that updated, let's just check that everything is still functional:

```
$ gradle build
...
BUILD SUCCESSFUL
```

One last piece of housekeeping before we start coding—let's copy the Secor Sequence-File that we downloaded from S3 at the end of section 7.3 to a data subfolder:

```
$ mkdir data
$ cp ../1_0_00000000000000000000 data/
```

Okay, good. Now let's write some Scala code! Still in the project root, start up the Scala console or REPL with this command:

```
$ gradle repl --console plain
...
scala>
```

Note that if you have Spark already installed on your local computer, you might want to use spark-shell instead.[6] Before we write any code, let's pull in imports that we

---

[6]  The Spark shell is a powerful tool to analyze data interactively: https://spark.apache.org/docs/latest/quick-start.html#interactive-analysis-with-the-spark-shell.

need. Type the following into the Scala console (we have omitted the `scala>` prompt and the console's responses for simplicity):

```
import org.apache.spark.{SparkContext, SparkConf}
import SparkContext._
import org.apache.spark.sql._
import functions._
import org.apache.hadoop.io.BytesWritable
```

Next, we need to create a `SparkConf`, which configures the kind of processing environment we want our job to run in. Paste this into your console:

```
val spark = SparkSession.builder()
  .appName("ShopperAnalysis")
  .master("local")
  .getOrCreate()
```

Feeding this configuration into a new `SparkContext` will cause Spark to boot up in our console:

```
scala> val sparkContext = spark.sparkContext
...
sparkContext: org.apache.spark.SparkContext =
 org.apache.spark.SparkContext@3d873a97
```

Next, we need to load the events file archived to S3 by Secor. Assuming this is in your data subfolder, you can load it like this:

```
scala> val file = sparkContext.
  sequenceFile[Long, BytesWritable](". /data/1_0_00000000000000000000")
...
file:org.apache.spark.rdd.RDD[(Long, org.apache.hadoop.io.BytesWritable)]
 = MapPartitionsRDD[1] at sequenceFile at <console>:23
```

Remember that a Hadoop SequenceFile is a binary key-value file format: in Secor's case, the key is a `Long` number, and the value is our event's JSON converted to a `BytesWritable`, which is a Hadoop-friendly byte array. The type of our `file` variable is now an `RDD[(Long, BytesWritable)]`; RDD is a Spark term, short for a *resilient distributed dataset*. You can think of the RDD as a way of representing the structure of your distributed collection of items at that point in the processing pipeline.

This `RDD[(Long, BytesWritable)]` is a good start, but really we just want to be working with human-readable JSONs in `String` form. Fortunately, we can convert the RDD to exactly this by applying this function to each of the values, by *mapping* over the collection:

```
scala> val jsons = file.map { case (_, bw) =>
  new String(bw.getBytes, 0, bw.getLength, "UTF-8")
}
jsons: org.apache.spark.rdd.RDD[String] = MapPartitionsRDD[2] at map
 at <console>:21
```

Note the return type: it's now an RDD[String], which is what we wanted. We're almost ready to write our aggregation code! Spark has a dedicated module, called Spark SQL (https://spark.apache.org/sql/), which we can use to analyze structured data such as an event stream. Spark SQL is a little confusingly named—we can use it without writing any SQL. First, we have to create a new SqlContext from our existing SparkContext:

```
val sqlContext = spark.sqlContext
```

A Spark SQLContext has a method on it to create a JSON-flavored data structure from a vanilla RDD, so let's employ that next:

```
scala> val events = sqlContext.read.json(jsons)
18/11/05 07:45:53 INFO FileInputFormat: Total input paths to process : 1
...
18/11/05 07:45:56 INFO DAGScheduler: Job 0 finished: json at <console>:24,
  took 0.376407 s
events: org.apache.spark.sql.DataFrame = [context: struct<timestamp:
  string>, directObject: struct<item: struct<price: double, product: string
  ... 1 more field>, order: struct<id: string, items:
  array<struct<product:string,quantity:bigint>> ... 1 more field>> ... 3
  more fields]
```

That's interesting—Spark SQL has processed all of the JSON objects in our RDD[String] and automatically created a data structure containing all of the properties it found! Note too that the output structure is now something called a DataFrame, not an RDD. If you have ever worked with the R programming language or with the Pandas data analytics library for Python, you will be familiar with data frames: they represent a collection of data organized into named columns. Where Spark-proper still leans heavily on the RDD type, Spark SQL embraces the new DataFrame type.[7]

We are now ready to write our aggregation code. To make the code more readable, we're going to define some aliases in Scala:

```
val (shopper, item, order) =
  ("subject.shopper", "directObject.item", "directObject.order")
```

These three aliases give us more convenient short-forms to refer to the three entities in our event JSON objects that we care about. Note that Spark SQL provides a dot-operator syntax to let us access a JSON property that is inside another property (as, for example, the *shopper* is inside the *subject*).

Now let's run our aggregation code:

```
scala> events.
  filter(s"${shopper} is not null").
  groupBy(shopper).
```

---

[7]  You can read more about this topic in "Introducing DataFrames in Apache Spark for Large-Scale Data Science," by Reynold Xin et al.: https://databricks.com/blog/2015/02/17/introducing-dataframes-in-spark-for-large-scale-data-science.html.

```
agg(
  col(shopper),
  sum(s"${item}.quantity"),
  sum(col(s"${item}.quantity") * col(s"${item}.price")),
  count(order),
  sum(s"${order}.value")
).collect
18/11/05 08:00:58 INFO CodeGenerator: Code generated in 14.294398 ms
...
18/11/05 08:01:07 INFO DAGScheduler: Job 1 finished: collect at
 <console>:43, took 1.315203 s
res0: Array[org.apache.spark.sql.Row] =
 Array([789,789,13,523.87,1,511.93], [456,456,1,1099.99,0,null],
 [123,123,6,319.94,1,179.97])
```

The `collect` at the end of our code forces Spark to *evaluate* our RDD and output the
result of our aggregations. As you can see, the result contains three rows of data, in a
slightly reader-hostile format. The three rows all match the table format depicted in
figure 7.12: the cells consist of a shopper ID, add-to-basket items and value, and finally
the number of orders and order value.

A few notes on the aggregation code itself:

- We filter out events with no shopper ID. This is needed because the Spark
  `SequenceFile` loader returns the file's empty header row, as well as our seven
  valid event JSON objects.
- We group by the shopper ID, and include that field in our results.
- We compose our aggregations out of various Spark SQL helper functions, including `col()` for column name, `sum()` for a sum of values across rows, and `count()`
  for a count of rows.

And that completes our experiments in writing a Spark job at the Scala console! You
have seen that we can build a sophisticated report for the Nile analytics team by using
Spark SQL. But running this inside a Scala console isn't a realistic option for the long-
term, so in the next section we will look briefly at operationalizing this code by using
Amazon's Elastic MapReduce platform.

### 7.4.4 *Running our job on Elastic MapReduce*

Setting up and maintaining a cluster of servers for batch processing is a major effort,
and not everybody has the need or budget for an always-running (persistent) cluster.
For example, if Nile wants only a daily refresh of the shopper spend analysis, we could
easily achieve this with a temporary (transient) cluster that spins up at dawn each day,
runs the job, writes the results to Amazon S3, and shuts down. Various data-processing-
as-as-a-service offerings have emerged to meet these requirements, including Amazon
Elastic MapReduce (EMR), Quobole, and Databricks Cloud.

Given that we already have an AWS account, we will use EMR to operationalize our
job in this section. But before we can productionize our job, we first need to consolidate

all of our code from the Scala console into a standalone Scala file. Copy the code from the following listing and add it into this file:

```
src/main/scala/nile/ShopperAnalysisJob.scala
```

The code in ShopperAnalysisJob.scala is functionally equivalent to the code we ran in the previous section. The main differences are as follows:

- We have improved readability a little (for example, by moving the byte wrangling for Secor's SequenceFile format into a dedicated function, toJson).
- We have created a main, ready for Elastic MapReduce to call, and we are passing in arguments to specify the input file and the output folder.
- Our SparkConf looks somewhat different; these are the settings required to run the job in a distributed fashion on EMR.
- Instead of running collect as before, we are now using the saveAsTextFile method to write our results back into the output folder.

**Listing 7.4   ShopperAnalysisJob.scala**

```scala
package nile

import org.apache.spark.{SparkContext, SparkConf}
import SparkContext._
import org.apache.spark.sql._
import functions._
import org.apache.hadoop.io.BytesWritable

object ShopperAnalysisJob {

  def main(args: Array[String]) {              // Our job expects the input
                                               // file and output folder to be
    val (inFile, outFolder) = {          ◁──── // supplied as arguments.
      val a = args.toList
      (a(0), a(1))
    }
                                                          // Ensures that Spark
    val sparkConf = new SparkConf()                       // will distribute the jar
      .setAppName("ShopperAnalysis")                      // file containing this job
      .setJars(List(SparkContext.jarOfObject(this).get))  ◁── // to each worker node
    val spark = SparkSession.builder()
      .config(sparkConf)
      .getOrCreate()
    val sparkContext = spark.sparkContext

    val file = sparkContext.sequenceFile[Long, BytesWritable](inFile)
    val jsons = file.map {
      case (_, bw) => toJson(bw)
    }

    val sqlContext = spark.sqlContext
    val events = sqlContext.read.json(jsons)
```

```
val (shopper, item, order) =
  ("subject.shopper", "directObject.item", "directObject.order")
val analysis = events
  .filter(s"${shopper} is not null")
  .groupBy(shopper)
  .agg(
    col(shopper),
    sum(s"${item}.quantity"),
    sum(col(s"${item}.quantity") * col(s"${item}.price")),
    count(order),
    sum(s"${order}.value")
  )

  analysis.rdd.saveAsTextFile(outFolder)     ◁──┐  Outputs our aggregates
}                                                  to a text file

  private def toJson(bytes: BytesWritable): String =   ◁──┐  Our byte wrangling for
    new String(bytes.getBytes, 0, bytes.getLength, "UTF-8")   Secor's SequenceFiles
}                                                              now lives in its own
                                                               function.
```

Now we are ready to assemble our Spark job into a *fat jar*—in fact, this jar is not so fat, as the only dependency we need to bundle into the jar is the Scala standard library; the Spark dependencies are already available on Elastic MapReduce, which is why we flagged those dependencies as provided. Build the fat jar from the command line like so:

```
$ gradle jar
:compileJava UP-TO-DATE
:compileScala UP-TO-DATE
:processResources UP-TO-DATE
:classes UP-TO-DATE
:jar

BUILD SUCCESSFUL

Total time: 3 mins 40.261 secs
```

The fat jar should now be available in our build subfolder:

```
$ file build/libs/spark-0.1.0.jar
build/libs/spark-0.1.0.jar: Zip archive data, at least v1.0 to extract
```

To run this job on Elastic MapReduce, we first have to make the fat jar available to the EMR platform. This is easy to do; we can simply upload the file to a new folder, jar, in our existing S3 bucket:

```
$ aws s3 cp build/libs/spark-0.1.0.jar s3://ulp-ch07-archive-${PET}/jar/ \
  --profile=ulp
upload: build/libs/spark-0.1.0.jar to s3://ulp-ch07-archive-little-
torty/jar/spark-0.1.0.jar
```

Before we can run the job, we need to log back into the AWS as our root user and assign the ulp user full administrator permissions; this is because our ulp user will need wide-ranging permissions in order to prepare the account for running EMR jobs. From the AWS dashboard:

1   Click the Identity & Access Management icon.
2   Click Users in the left-hand navigation pane.
3   Click your ulp user.
4   Click the Add Permissions button.
5   Click the Attach Existing Policies Directly tab.
6   Select the AdministratorAccess policy and click Next: Review.
7   Click the Add Permissions button.

Before we can run our job, we need to create an EC2 keypair plus IAM security roles for Elastic MapReduce to use. From inside your virtual machine, enter the following:

```
$ aws emr create-default-roles --profile=ulp --region=eu-west-1
$ aws ec2 create-key-pair --key-name=spark-kp --profile=ulp \
  --region=eu-west-1
```

This will create the new security roles required by Elastic MapReduce to run a job. Now run this:

```
$ BUCKET=s3://ulp-ch07-archive-${PET}
$ IN=${BUCKET}/secor_dev/backup/raw-events-
➥ ch07/offset=0/1_0_00000000000000000000
$ OUT=${BUCKET}/out
$ aws emr create-cluster --name Ch07-Spark --ami-version 3.6 \
--instance-type=m3.xlarge --instance-count 3 --applications Name=Hive \
--use-default-roles --ec2-attributes KeyName=spark-kp \
--log-uri ${BUCKET}/log --bootstrap-action \
➥ Name=Spark,Path=s3://support.elasticmapreduce/spark/install-spark,
➥ Args=[-x] --steps Name=ShopperAnalysisJob,Jar=s3://eu-west-1.
➥ elasticmapreduce/libs/script-runner/script-runner.jar,
➥ Args=[/home/hadoop/spark/bin/spark-submit,--deploy-mode,cluster,
➥ --master,yarn-cluster,--class,nile.ShopperAnalysisJob,
${BUCKET}/jar/spark-0.1.0.jar,${IN},${OUT}] \
--auto-terminate --profile=ulp --region=eu-west-1
{
    "ClusterId": "j-2SIN23GBVJ0VM"
}
```

The last three lines—the JSON containing the cluster ID—tell us that the cluster is now starting up. Before we take a look at the cluster, let's break down the preceding create-cluster command into its constituent parts:

1   Start an Elastic MapReduce cluster called Ch07-Spark using the default EMR roles.
2   Start up three m3.xlarge instances (one master and two slaves) running AMI version 3.6.

3   Log everything to a log subfolder in our bucket.

4   Install Hive and Spark onto the cluster.

5   Add a single job step, `nile.ShopperAnalysisJob`, as a Spark job found inside the jar/spark-0.1.0.jar file in our bucket.

6   Provide the in-file and out-folder as first and second arguments to our job, respectively.

7   Terminate the cluster when the Spark job step is completed.

A bit of a mouthful! In any case, we can now go and watch our cluster starting up. From the AWS dashboard:

1   Make sure you have the Ireland region selected at the top right (or the region you've specified when provisioning the cluster).

2   In the Analytics section, click EMR.

3   In your Cluster List, click the job called Ch07-Spark.

You should see EMR first provisioning the cluster, and then bootstrapping the servers with the required software, as per figure 7.13.

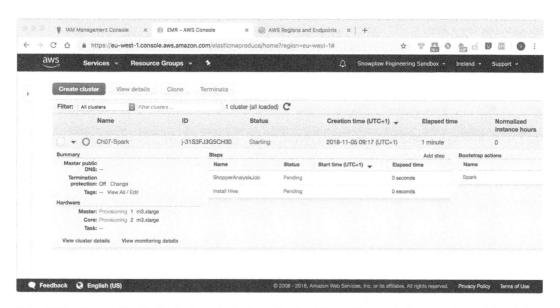

**Figure 7.13   Our new Elastic MapReduce cluster is running bootstrap actions on both our master and slave (aka core) instances.**

Wait a little while, and the job status should change to Running. At this point, scroll down to the Steps subsection and expand both job steps, Install Hive and ShopperAnalysisJob. You can now watch these steps running, as shown in figure 7.14.

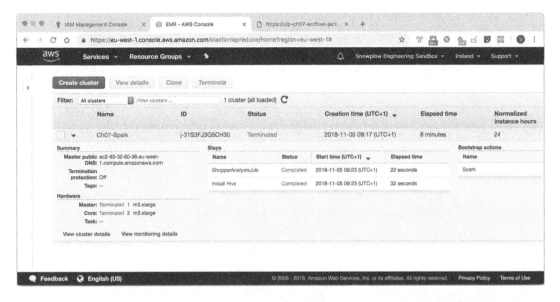

**Figure 7.14  Our cluster has successfully completed running two job steps: the first step installed Hive on the cluster, while the second ran our Spark ShopperAnalysisJob. This is a helpful screen for debugging any failures in the operation of our steps.**

If everything has been set up correctly, each step's status should move to Completed, and then the overall cluster's status should move to Terminated, All Steps Completed. We can now admire our handiwork:

```
$ aws s3 ls ${OUT}/ --profile=ulp
2018-11-05 08:22:35            0 _SUCCESS
2018-11-05 08:22:30            0 part-00000
...
2018-11-05 08:22:30           25 part-00017
...
2018-11-05 08:22:31           23 part-00038
...
2018-11-05 08:22:31           24 part-00059
...
2018-11-05 08:22:35            0 part-00199
```

The _SUCCESS is a slightly old-school *flag file*: an empty file whose arrival tells any downstream process that is monitoring this folder that no more files will be written to this folder. The more interesting output is our part- files. Collectively, these files represent the output to our Spark job. Let's download them and review the contents:

```
$ aws s3 cp ${OUT}/ ./data/ --recursive --profile=ulp
...
download: s3://ulp-ch09-archive-little-torty/out/_SUCCESS to data/_SUCCESS
...
```

```
download: s3://ulp-ch09-archive-little-torty/out/part-00196 to
 data/part-00196
$ cat ./data/part-00*
[789,13,499.99,1,511.93]
[456,1,1099.99,0,null]
[123,6,319.94,1,179.97]
```

And there are our results! Don't worry about the number of empty part- files created; that is an artifact of the way Spark is dividing its processing work into smaller work units. The important thing is that we have managed to transfer our Spark job to running on a remote, transient cluster of servers, completely automated for us by Elastic MapReduce.

If Nile were a real company, the next step for us would be to automate the operation of this job further, potentially using the following:

- A library for running and monitoring the job, such as boto (Python), Elasticity (Ruby), Spark Plug (Scala), or Lemur (Clojure)
- A tool for scheduling the job to run overnight, such as cron, Jenkins, or Chronos

We leave those as exercises to you! The important thing is that you have seen how to develop a batch processing job locally by using the Scala console/REPL, and then how to put that job into operation on a remote server cluster using Elastic MapReduce. Almost everything else in batch processing is just a variation on this theme.

## Summary

- A unified log such as Apache Kafka or Amazon Kinesis is not designed as a long-term store of events. Kinesis has a hard limit, or trim horizon, of a maximum of 168 hours, after which records are culled from a stream.
- Archiving our events into long-term storage enables three important requirements: reprocessing, robustness, and refinement.
- Event *reprocessing* from the archive is necessary when additional analytics requirements are identified, or bugs are found in existing processing.
- Archiving all raw events gives us a *more robust* pipeline. If our stream processing fails, we have not lost our event stream.
- An event archive can be used to *refine* our stream processing. We can improve the accuracy of our processing in the face of late-arriving data, deliberate approximations we applied for performance reasons, or inherent limitations of our stream processing framework.
- We should archive our rawest events (as upstream in our topology as possible) to a distributed filesystem such as HDFS or Amazon S3, using a tool such as Pinterest's Secor, Snowplow's kinesis-s3, or Confluent's Connect S3.
- Once we have our events archived in a distributed filesystem, we can perform processing on those events by using a batch processing framework such as Apache Hadoop or Apache Spark.

- Using a framework such as Spark, we can write and test an event processing job interactively from the Scala console or REPL.
- We can package our Spark code as a fat jar and run it in a noninteractive fashion on a hosted batch-processing platform such as Amazon Elastic MapReduce (EMR).

# Railway-oriented processing

*8*

**This chapter covers**

- Handling failure within Unix programs, Java exceptions, and error logging
- Designing for failure inside and across stream processing applications
- Composing failures inside work steps with the Scalaz `Validation`
- Failing fast across work step boundaries with Scala's `map` and `flatMap`

So far, we have focused on what you might call the *happy path* within our unified log. On the happy path, events successfully validate against their schemas, inputs are never accidentally null, and Java exceptions are so rare that we don't mind them crashing our application.

The problem with focusing exclusively on the happy path is that failures *do* happen. More than this: if you implement a unified log across your department or company, failures will happen *extremely frequently*, because of the sheer volume of events flowing through, and the complexity of your event stream processing. Linus's

171

law states, "Given enough eyeballs, all bugs are shallow."[1] Adapting this, a law of unified log processing might be as follows:

*Given enough events, all bugs are inevitable.*

If we can expect and design for inevitable failure inside our stream processing applications, we can build a much more robust unified log, one that hopefully won't page us regularly at 2 a.m. because it crashed Yet Another `NullPointerException` (YANPE). This chapter, then, is all about designing for failure, using an overarching approach that we will call *railway-oriented processing*. We have been using this approach at Snowplow throughout our event pipeline since the start of 2013, so we know that it can enable the processing of billions of events daily with a minimum of disruption.

For reasons that should become clear as we delve deeper into this topic, this chapter will be the first one where we work predominantly in Scala. Scala is a strongly typed, hybrid object-oriented and functional language that runs on the Java virtual machine. Scala has great support for railway-oriented processing—capabilities that even Java 8 lacks.

So, all aboard the railway-oriented programming express, and let's get started.

## 8.1    Leaving the happy path

> *Two roads diverged in a wood, and I—*
> *I took the one less traveled by,*
> *And that has made all the difference.*
>
> —Robert Frost, "The Road Not Taken" (1916)

Before diving into railway-oriented processing, let's look at how failure is handled in two distinct environments that many readers will be familiar with: Unix programs and Java. We'll then follow this up with general thoughts on error logging as it is practiced today.

### 8.1.1    Failure and Unix programs

Unix is designed around the idea that failures will happen. Any process that runs in a Unix shell will return an exit code when it finishes. The convention is to return zero for success, and an integer value higher than zero in the case of failure.

This exit, or return, code isn't the only communication channel available to a Unix program; each program also has access to three standard streams (aka I/O file descriptors): `stdin`, `stdout`, and `stderr`. Table 8.1 provides the properties of these three streams.

---

[1]  Linus's law is further explained at Wikipedia, https://en.wikipedia.org/wiki/Linus%27s_Law.

**Table 8.1    The three standard streams supported by Unix programs**

| Short name | Long name | File descriptor | Description |
|---|---|---|---|
| stdin | Standard in | 0 | The input stream of data going into a Unix program |
| stdout | Standard out | 1 | The output stream where a Unix program writes data related to successful operation |
| stderr | Standard error | 2 | The output stream where a Unix program writes data related to failed operation |

Putting the exit codes and the three standard streams together, we can represent a Unix program's happy and failure paths, as shown in figure 8.1.

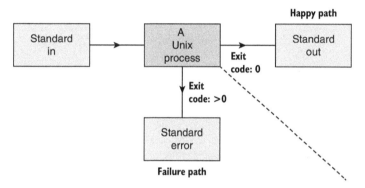

**Figure 8.1    A Unix program reads from standard in and can respond with exit codes and output streams along a happy path or a failure path.**

It's only correct to note, however, that things are not always as clear-cut as figure 8.1 implies:

- A Unix process that ends up failing with a nonzero exit code may well also write output to stdin before it fails.
- Likewise, a chatty Unix process may write warnings or diagnostic output to stderr before ultimately returning with a zero code indicating success.

How *composable* is failure handling in Unix programs? By *composable*, we mean that can we combine multiple successes and failures, and the combined result will still make sense. Let's see—if you have a Unix terminal handy, type in the following:

```
$ false | false | true; echo $?
0
```

If you are not familiar with false and true, these are super-simple Unix programs that return exit codes of 1 (failure) and 0 (success), respectively. In this example, we are combining two false programs with a true program, using the vertical bar character | (also known as a *pipe*) to make a *Unix pipeline*.

As you can see, the first two failures do not cause the pipeline to fail, and the ultimate return code of the pipeline is the return code of the last program run.

In some Unix shells (ksh, zsh, bash), we can do a little better than this by using the built-in `pipefail` option:

```
$ alias info='>&2 echo'
$ set -o pipefail; info s1 | false | info s3 | true; echo $?
s3
s1
1
```

The `pipefail` option sets the exit code to the exit code of the last program to exit nonzero, or returns 0 if all exited successfully. This is an improvement, but the s3 output shows that the pipeline is still not short-circuiting, or failing fast, after an individual component within it fails. This is because a Unix pipeline is chaining input and output streams together, *not* exit codes, as demonstrated in figure 8.2.

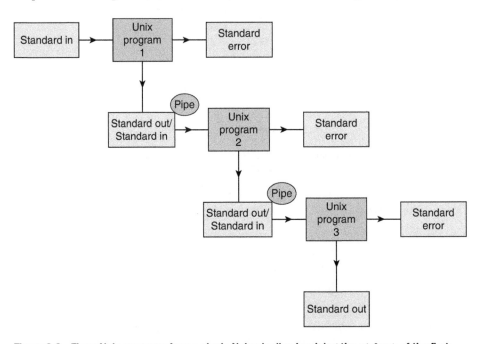

**Figure 8.2  Three Unix programs form a single Unix pipeline by piping the `stdout` of the first program into the `stdin` of the next program. Many shells support an option to pipe both `stdout` and `stderr` into `stdin`, but in both cases the program's exit codes are ignored.**

If we do want to fail fast, we have to put our commands in a shell script that uses the `set -e` option, which will terminate the script as soon as any command within the script returns a nonzero error code. If you run the code in the following listing, you will see the following output; notice that the second echo in the shell script is never reached:

```
$ ./fail-fast.bash; echo $?
s1
1
```

Listing 8.1   fail-fast.bash

```
#!/bin/bash
set -e

echo "s1"
false                    This line is never
echo "s3"          ◁─────┘ reached.
```

In sum: Unix programs and the Unix pipeline have powerful yet easy-to-understand features for dealing with failure. But they are not as composable as we would like.

### 8.1.2   Failure and Java

Let's take a look now at how Java deals with failure. Java depends heavily on exceptions for handling failures, making a distinction between two types of exceptions:

- *Unchecked exceptions*—RuntimeException, Error, and their subclasses. Unchecked exceptions represent bugs in your code that a caller cannot be expected to recover from: the dread NullPointerException is an unchecked exception.
- *Checked exceptions*—Exception and its subclasses, except RuntimeException (which, strangely, is a subclass of Exception). In Java, every method must declare any uncaught checked exceptions that can be thrown within its scope, passing the responsibility on to the caller to handle them as they decide.

Let's look at the following HelloCalculator app, where we employ both unchecked and checked exceptions. The following listing contains the HelloCalculator.java code, annotated to show the use of both exception types.

Listing 8.2   HelloCalculator.java

```
package hellocalculator;

import java.util.Arrays;

public class HelloCalculator {

    public static void main(String[] args) {
        if (args.length < 2) {
            String err = "too few inputs (" + args.length + ")";
            throw new IllegalArgumentException(err);
        } else {
            try {
                Integer sum = sum(args);
                System.out.println("SUM: " + sum);
            } catch (NumberFormatException nfe) {
                String err = "not all inputs parseable to Integers";
                throw new IllegalArgumentException(err);
```

**NumberFormat-Exception is a checked Exception, so we have to catch it here.**

**IllegalArgument-Exception is an unchecked RuntimeException, which will cause our program to terminate.**

```
        }
      }
    }
```

**This method could throw a NumberFormatException, but we don't catch it, so we have to declare it as part of our method's API.**

```
static Integer sum(String[] args) throws NumberFormatException {    ⊲─┘
  return Arrays.asList(args)
    .stream()
    .mapToInt(str -> Integer.parseInt(str))    ⊲─┐
    .sum();                                       │  Integer.parseInt's own API
}                                                 │  declares that it could throw
                                                  │  a NumberFormatException.
}
```

Java's built-in failure handling is broadly designed around two failure scenarios:

- We have an unrecoverable bug and we want to terminate the program.
- We have a potentially recoverable issue and we want to try to recover from it (and if we can't recover from it, we terminate the program).

But what if we cannot recover from a failure but *don't* want to terminate our overall program? This might sound counterintuitive: how can our program keep going with an *unrecoverable failure*? To be sure, many programs cannot—but some can, especially if they consist of processing many much smaller *units of work*, for example:

- A web server, responding to thousands of HTTP requests a minute. Each request-response exchange is a unit of work.
- A stream processing job in our unified log, tasked with enriching many millions of individual events. The enrichment of each incoming event is a unit of work.
- A web-scraping bot, parsing thousands of web pages to look for product price information. The parsing of each web page is a unit of work.

In each of these scenarios, the programmer may prefer to route the unrecoverable unit of work to a failure path rather than terminate the whole program, as shown in figure 8.3.

As with many other languages, Java does not have any built-in tools for routing failing units of work to a failure path. In situations like these, a Java programmer will often fall back to simply logging the failure as an error and skipping to the next unit of work. To demonstrate this, let's imagine an updated version of the original Hello-Calculator code: instead of summing all arguments together, our new version will increment (add 1 to) any numeric arguments, and log an error message for any arguments that are not numeric. This is illustrated in the following listing.

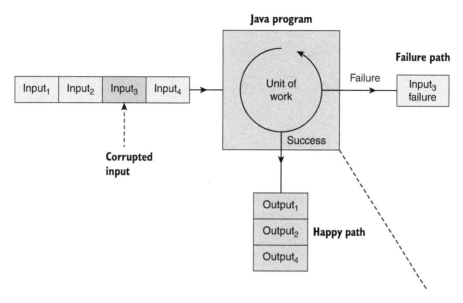

**Figure 8.3** Our Java program is processing four items of input, one of which is somehow corrupted. Our Java program performs the unit of work on each of the four inputs: three are successfully output along the happy path, but the corrupted input throws an error and ends up on the failure path.

---

**Listing 8.3   HelloIncrementor.java**

```java
package hellocalculator;

import java.util.Arrays;
import java.util.logging.Logger;

public class HelloIncrementor {

  private final static Logger LOGGER =
    Logger.getLogger(HelloIncrementor.class.getName());

  public static void main(String[] args) {
    Arrays.asList(args)
      .stream()
      .forEach((arg) -> {
        try {
          Integer i = incr(arg);
          System.out.println("INCREMENTED TO: " + i);
        } catch (NumberFormatException nfe) {
          String err = "input not parseable to Integer: " + arg;
          LOGGER.severe(err);
        }
      });
  }
```

We use the built-in java.util.Logging for our error logging.

Loop through all command-line arguments.

If we can increment, print out the incremented value.

If we cannot increment, log the error at the highest log level.

```
static Integer incr(String s) throws NumberFormatException {
  return Integer.parseInt(s) + 1;
}
}
```

Copy the code from listing 8.3 and save it into a file, like so:

```
hellocalculator/HelloIncrementor.java
```

Next, we will compile our code and run it, supplying three valid and one invalid (non-numeric) argument:

```
$ javac hellocalculator/HelloIncrementor.java
$ java hellocalculator.HelloIncrementor 23 52 a 1
INCREMENTED TO: 24
INCREMENTED TO: 53
Nov 13, 2018 5:42:30 PM hellocalculator.HelloIncrementor lambda$main$0
GRAVE: input not parseable to Integer: a
INCREMENTED TO: 2
```

See how our failure is buried in the middle of our successes? We have to look carefully at the program's output to discern the failure output.

Worse, we have just outsourced this program's failure path to our logging framework to handle. We can say that the failure path is now *out-of-band*: the failures are no longer present in the source code or influencing the program's *control flow*. Incidentally, many experienced programmers criticize the concept of exceptions for a similar reason: they create a secondary control flow in a program, one that exists outside the standard imperative flow, and thus is difficult to reason about.[2]

Whatever the criticisms of exceptions, error logging is surely worse: it is not just outside the program's main control flow, but also outside the program itself. The challenges of dealing with out-of-band error logging have fueled a whole software industry, which we will briefly look at next.

### 8.1.3   *Failure and the log-industrial complex*

A Java program, a Node.js program, and a Ruby program all walk into a bar. Each logs an error on the way in. The joke is on the barman, because they are all speaking different languages:

```
ERROR  2018-11-13 06:50:14,125 [Log_main]   "com.acme.Log": Error from Java
Error from node.js
E, [851000 #0] ERROR -- : Error from Ruby
```

Amazingly, no common format exists for program logging: different languages and frameworks all have their own logging levels (error, warning, and so forth) and log message formats. Combine this with the fact that log messages are written using

---

[2] Joel Spolsky's essay on why exceptions are not always a good thing: https://www.joelonsoftware.com/2003/10/13/13/

human language, and you are left with logs that can often be analyzed using only plain-text search.

Various logistical issues also are associated with outsourcing your failure path to a logging framework:

- What if we run out of space on our server to store logs?
- In a world of transient virtual servers and ephemeral stateless containers, how do we ensure that we collect our logs before the server itself is terminated?
- How do we collect logs from client devices that we don't control?

Collectively dealing with these issues around logging have spawned what we might call, only half-jokingly, the *log-industrial complex*, consisting of the following:

- *Logging frameworks and facades*—Java alone has Log4j, SLF4J, Logback, java.util .Logging, tinylog, and others.
- *Log collection agents and frameworks*—These include Apache Flume, Logstash, Filebeat, Facebook's now-shuttered Scribe project, and Fluentd.
- *Log storage and analytics tools*—These include Splunk, Elasticsearch plus Kibana, and Sawmill.
- *Error collection services*—These include Sentry, Airbrake, and Rollbar. These services are often focused on collecting errors from client devices.

Even with all of this tooling, in a unified log context we still have another unsolved problem: how can we reprocess a failed unit of work (for example, an event) if and when the underlying issue that caused the failure is fixed?

In a unified log world, there has to be a better way of working with our failure path. We will explore this next.

## 8.2 Failure and the unified log

*Reports that say that something hasn't happened are always interesting to me, because as we know, there are known knowns; there are things we know we know. We also know there are known unknowns; that is to say, we know there are some things we do not know. But there are also unknown unknowns—the ones we don't know we don't know.*

—US Secretary of Defense Donald H. Rumsfeld (February 12, 2002)

In this section, we will set out a simple pattern for unified log processing that accounts for the failure path as well as the happy path. For our treatment of the failure path, we will borrow from the better ideas of section 8.1 and throw in some new ideas of our own.

### 8.2.1 A design for failure

How should we be handling failure in our unified log processing? First, let's propose some rules governing *program termination*. We should terminate our job only if one of the following occurs:

- We encounter an unrecoverable error during the initialization phase of our job.
- We encounter a novel error while processing a unit of work inside our job.

A *novel error* means one that we haven't seen before: a Rumsfeld-esque unknown unknown. Although it might be tempting to keep processing in this case to minimize disruption, terminating the job forces us to evaluate any novel error as soon as we encounter it. We can then determine how this new error should be handled in the future—for example, can it be recovered from without failing the unit of work?

Next, how do we handle an unrecoverable but not-unexpected error within a unit of work? Here there is no way around it—we have to move this unit of work into our failure path, but making sure to follow a few important rules:

- Our failure path must not be out-of-band. We don't need to rely on third-party logging tools; we are implementing a unified log, so let's use it!
- Entries in our failure path must contain the reason or reasons for the failure in a well-structured form that both humans and machines can read.
- Entries in our failure path must contain the original input data (for example, the processed event), so that the unit of work can potentially be replayed if and when the underlying issue can be fixed.

Let's make this a little more concrete with the example of a simple stream-processing job that is enriching individual events, as shown in figure 8.4.

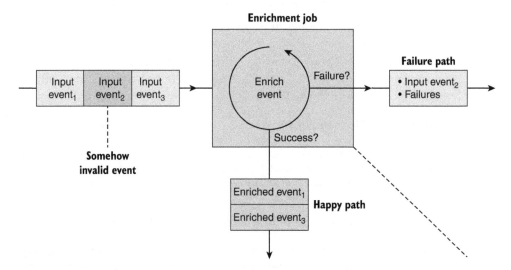

**Figure 8.4   Our enrichment job processes events from the input event stream, writes enriched events to our happy path event stream, and writes input events that failed enrichment, plus the reasons why they failed enrichment, to our failure path event stream.**

Does the flow in figure 8.4 look familiar? It shares a lot in common with Unix's concept of three standard streams: our stream processing app will *read* from one event stream, and *write* to two event streams, one for the happy path and the other for individual failures. At the same time, we have improved some of the failure-handling techniques of section 8.1:

- *We have done away with exit values.* The success or failure of a given unit of work is reflected in whether output was written to the happy event stream or the failure event stream.
- *We have removed any ambiguity in the outputs.* A unit of work results in output *either* to the happy stream or to the failure stream, never both, nor neither. Three input events mean a total of three output events.
- *We are using the same in-band tools to work with both our successes and our failures.* Failures will end up as well-structured entries in one event stream; successes will end up as well-structured entries in the other event stream.

This is a great start, but what do we mean when we say that our failures will end up as "well-structured entries" in an event stream? Let's explore this next.

## 8.2.2 Modeling failures as events

How could we describe our stream processing job's failure to enrich an event read from an input stream? Perhaps something like this:

> *At 12:24:07 in our production environment, SimpleEnrich v1 failed to enrich Inbound Event 428 because it failed JSON Schema validation.*

Does this look familiar? It contains all the same grammatical components as the events we introduced in chapter 2:

- *Subject*—In this case, our stream processing job, SimpleEnrich v1, is the entity carrying out the action of this event.
- *Verb*—The action being done by the *subject* is, in this case, "failed to enrich."
- *Direct object*—The entity to which the action is being done is Inbound Event 428.
- *Timestamp*—This tells us exactly when this failure occurred.

We have another piece of *context* besides the timestamp: the environment in which the failure happened. Finally, we also have a *prepositional object*—namely, the reason that the enrichment failed. We call this *prepositional* because it is associated with the event via a prepositional phrase—in this case, "because of." Figure 8.5 clarifies the relationship between our event's constituent parts: *subject, verb, direct object, prepositional object* (the reason for the failure), and *context*.

It may seem a little strange to have the direct object of our failure event be an event—specifically, the inbound event that failed enrichment. But this "turtles all the way down" design is powerful: it means that our failure event contains within it all the information required to retry the failed processing in the future. Without this design, we would have to manually correlate the failure messages with the original input events if we wanted to attempt reprocessing of failed events in the future.

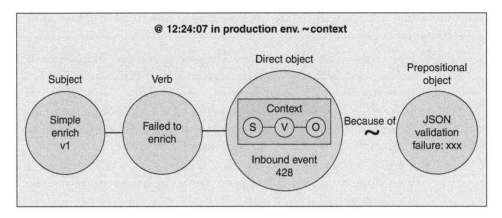

**Figure 8.5   We are representing our enrichment as an event: the subject is the enrichment job itself, the verb is "failed to enrich," and the direct object is the event we failed to enrich.**

This might sound a little theoretical; after all, the events have failed processing with an unrecoverable error once. What makes us think that this error might be recoverable in the future? In fact, there are lots of reasons that we shouldn't be binning this failed event just yet:

- Perhaps the enrichment failed because of a problem with a third-party service that has since been fixed. For example, the third-party service had an outage due to a denial-of-service attack, which has since been resolved.
- Perhaps some events failed enrichment because they were serialized with a version of a schema that somebody had forgotten to upload into our schema repository. Once this has been fixed and the schema uploaded, the events can be reprocessed.
- Perhaps a whole set of events failed enrichment because the source system had corrupted them all in a systemic way—for example, the source system had accidentally URL-encoded the customers' email addresses. We could write a quick stream processing job to read the corrupted events, fix them, and write them back to the original stream, ready for a second attempt at enrichment, as in figure 8.6.

This last example is starting to hint at a key aspect of this design: its *composability* across multiple event streams within our unified log. We'll look at this next.

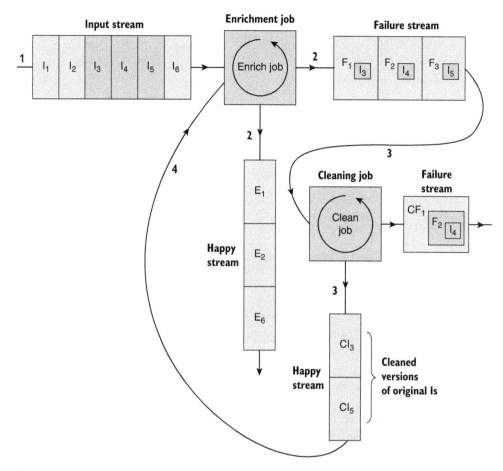

**Figure 8.6** Our enrichment job reads six input events; three fail enrichment and are written out to our failure stream. We then feed those three failures into a cleaning job that attempts to fix the corrupted events. The job manages to fix two events, which are fed back into the original enrichment job.

### 8.2.3 Composing our happy path across jobs

If our stream processing jobs follow the simple architecture set out previously, we can create a happy path composed of multiple stream processing jobs: the happy path output of one job is fed in as the input of the next job. Figure 8.7 illustrates this process.

Figure 8.7 deliberately elides how we handle the failure paths of our individual jobs. How we handle the failure events emitted by a given job will depend on a few factors:

- Do we expect to be able to recover from a job's failures in the future, and do we care enough to attempt recovery?
- How will we monitor error rates, and what constitutes an acceptable error rate for a given job?
- Where will we archive our failures?

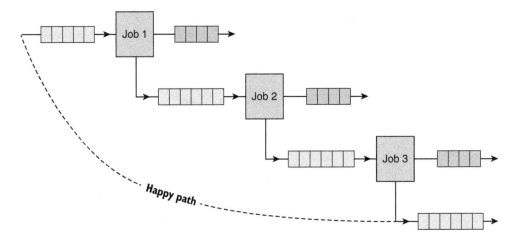

**Figure 8.7   We have composed a happy path by chaining together the happy output event stream of each stream processing job as the input stream of the next job.**

These kinds of questions may well have different answers for different stream-processing jobs. For example, a job auditing customer refunds might take individual failures much more seriously than a job providing vanity metrics for a display panel in reception. But by modeling our failures as events, we should be able to create highly reusable failure-handling jobs, because they speak our *lingua franca* of well-structured events.

## 8.3    *Failure composition with Scalaz*

> *If you fail to plan, you are planning to fail!*
>
> —Benjamin Franklin

So far, we have looked at the concept of the failure path at an architectural level, but how do we implement these patterns inside our stream processing jobs? It's time to add a new tag team to our unified log arsenal: Scala and Scalaz.

### 8.3.1   *Planning for failure*

Imagine that we are working at an online retailer that sells products in three currencies (euros, dollars, and pounds), but does all of this financial reporting in euros, its base currency. Our retailer already has all customer orders available in a stream in its unified log, so the managers want us to write a stream processing job that does the following:

- *Reads* customer orders from the incoming event stream
- *Converts* all customer orders into euros
- *Writes* the updated customer orders to a new event stream

The currency conversion will be done using the live (current) exchange rate, to keep things simple. Our job can look up the exchange rate by making an API call to a third-party service called Open Exchange Rates (https://openexchangerates.org/).

This sounds simple. What could go wrong with the operation of our job? In fact, a few things:

- Perhaps Open Exchange Rates is having an unplanned outage or planned downtime to facilitate an upgrade.
- Perhaps test data made it into the production flow, and the customer order is in a currency other than one of the three allowed ones.
- Perhaps a hacker has managed to transact an order with a non-numeric order value, getting themselves a huge flat screen TV for ¥l33t.

An interesting thing about these failures is that they are, as a management consultant might say, the opposite of *mutually exclusive, collectively exhaustive* (MECE):

- They are not *mutually exclusive*. We could have the order in the wrong currency and the order value be non-numeric.
- They are not *collectively exhaustive*. We could always experience a *novel error*. If we experience a novel error, we will add it to our failure handling, but there can always be novel errors.

Putting all this together, it is clear that our stream processing job will need to plan for failure: in addition to the output stream of orders in our base currency—our happy path—we will need a second output stream to report our currency conversion failures—our failure path. Figure 8.8 illustrates the overall job flow.

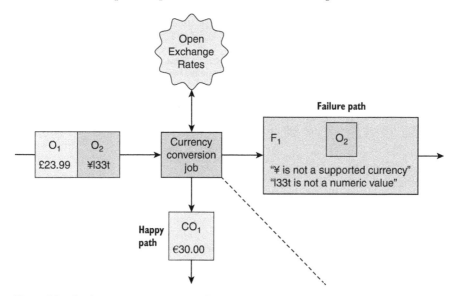

**Figure 8.8  Our input event stream contains two customer order events. The valid one is successfully converted into euros and written to our happy stream. The second event is invalid and is written to our failure stream; our failure event records the failures encountered as well as the original event.**

We are not particularly interested in the "plumbing" of this job; we have explored the mechanics of reading and writing to event streams in detail in previous chapters. This chapter is all about planning for failure—so the important thing is learning how to assemble the internal logic of this job in a way that can cope with all these possible failures.

For these purposes, we can work with a stripped-down Scala application—just a few functions, unit tests, and a simple command-line interface.

### 8.3.2   *Setting up our Scala project*

Let's get started. We are going to create our Scala application by using Gradle. Scala Build Tool (SBT) is the more popular build tool for Scala, but we are familiar with Gradle and it works fine with Scala, so let's stick with that.

First, create a directory called forex, and then switch to that directory and run the following:

```
$ gradle init --type scala-library
...
BUILD SUCCESSFUL
...
```

As we did in chapter 3, we'll now delete the stubbed Scala files that Gradle created:

```
$ rm -rf src/*/scala/*
```

The default build.gradle file in the project root isn't quite what we need either, so replace it with the following code.

#### Listing 8.4   build.gradle

```
apply plugin: 'scala'          ◁─┐  We use Gradle's Scala plugin for
                                 └─ projects involving Scala code.

repositories {
  mavenCentral()                      ┐ Some of our dependencies
  maven {                      ◁──────┘ are published to Sonatype.
    url 'http://oss.sonatype.org/content/repositories/releases'
  }
}

version = '0.1.0'                                       ┐ So that Gradle
                                                        │ can compile Scala
                                                        │ against Java 8
ScalaCompileOptions.metaClass.daemonServer = true  ◁───┘
ScalaCompileOptions.metaClass.fork = true
ScalaCompileOptions.metaClass.useAnt = true
ScalaCompileOptions.metaClass.useCompileDaemon = false
                                                       ┐ For our runtime, we
dependencies {                                     ◁───┤ need Scala and scalaz;
  runtime 'org.scala-lang:scala-compiler:2.12.7'       │ for testing, we will use
  compile 'org.scala-lang:scala-library:2.12.7'        ┘ Specs2.
  compile 'org.scalaz:scalaz-core_2.12:7.2.27'
  testCompile 'org.specs2:specs2_2.12:3.8.9'
```

```
    testCompile 'org.typelevel:scalaz-specs2_2.12:0.5.2'
}

task repl(type:JavaExec) {                          Task to start the
    main = "scala.tools.nsc.MainGenericRunner"      Scala console
    classpath = sourceSets.main.runtimeClasspath
    standardInput System.in
    args '-usejavacp'
}
```

With that updated, let's check that everything is still functional:

```
$ gradle build
...
BUILD SUCCESSFUL
...
```

Okay, good—now let's write some Scala code!

### 8.3.3 From Java to Scala

Remember that we need to check whether our customer order is one of our three supported currencies (euros, dollars, or pounds).

If we were still working in Java, we might write a function that threw a checked exception if the currency was not in one of our supported currencies. In the following listing, we have a function that parses the currency string:

- If it is a supported currency, the function returns our currency as a Java enum.
- Otherwise, it throws a checked exception, UnsupportedCurrencyException.

---

**Listing 8.5  CurrencyValidator.java**

```
package forex;

import java.util.Locale;
                                                    Defines our three
public class CurrencyValidator {                    supported currencies
    public enum Currency {USD, GPB, EUR}            as a Java enum

    public static class UnsupportedCurrencyException      A checked exception
        extends Exception {                               for unsupported
        public UnsupportedCurrencyException(String raw) { currencies
            super("Currency must be USD/EUR/GBP, not " + raw);
        }
    }
                                                        Our validation function
    public static Currency validateCurrency(String raw)  throws our checked
        throws UnsupportedCurrencyException {             exception if the currency
                                                          is unsupported.
        String rawUpper = raw.toUpperCase(Locale.ENGLISH);
        try {                                           We catch the unchecked
            return Currency.valueOf(rawUpper);          exception and convert it
        } catch (IllegalArgumentException iae) {        into our checked exception.
```

```
      throw new UnsupportedCurrencyException(raw);
    }
  }
}
```

One of the nice aspects of Scala for recovering Java programmers is that it's possible to port existing Java code to Scala code (albeit unidiomatic Scala code) with only cosmetic syntactical changes. The following listing contains just such a direct, unidiomatic port of our existing `CurrencyConverter` class to a Scala object.

**Listing 8.6   currency.scala**

```
package forex

object Currency extends Enumeration {          ◁──  Our Java enum is now a
  type Currency = Value                              Scala object extending
  val Usd = Value("USD")                             Enumeration.
  val Gbp = Value("GBP")
  val Eur = Value("EUR")                                          Our Java
}                                                          exception is now a
                                                          Scala case class.
case class UnsupportedCurrencyException(raw: String)
    extends Exception("Currency must be USD/EUR/GBP, not " + raw)   ◁──

object CurrencyValidator1 {                     ◁───────  Our Java class is
                                                          now a Scala object.
  @throws(classOf[UnsupportedCurrencyException])
  def validateCurrency(raw: String): Currency.Value = {

    val rawUpper = raw.toUpperCase(java.util.Locale.ENGLISH)
    try {
      Currency.withName(rawUpper)          ◁──────        Last expression
    } catch {                                             evaluated is returned
      case nsee: NoSuchElementException =>   ◁──┐         without needing return
        throw new UnsupportedCurrencyException(raw)        statement
    }
  }                          Catching exceptions in
}                            Scala uses pattern-
                             matching semantics.
```

If you are completely new to Scala, here are a few immediate things to note about this code compared to Java:

- A single Scala file can contain multiple independent classes and objects—in which case, we give the file a descriptive name starting with a lowercase letter.
- No need for semicolons at the end of each line, though they are still useful to separate multiple statements on the same line.
- We use `val` to assign what in Java would be variables. The internal state of a `val` can be modified, but a `val` cannot be reassigned, so no `val a = 0; a = a + 1`.
- In place of a class containing a static method, we now have a *Scala object* containing a method. An object in Scala is a singleton, a unique instance of a class.

- A function definition starts with the keyword def.
- Scala has type inference, which means that it has the ability to figure out the types that are left off.

There is an interesting difference in the behavior of our `validateCurrency` function too: our function now has the `@throws` annotation recording the exception that it may throw. Scala doesn't distinguish between checked and unchecked exceptions, so this annotation is optional, and useful only if some Java code is going to call this function.

Another nice difference between Scala and Java is the interactive console, or *read-eval-print loop* (REPL), available with Scala. Let's use this to put our new Scala code through its paces. From the project root, start up the REPL with this command:

```
$ gradle repl --console plain
...
scala>
```

The Scala prompt is waiting for you to type in a Scala statement to execute. Let's try this one:

```
scala> forex.CurrencyValidator1.validateCurrency("USD")
res0: forex.Currency.Value = USD
```

The second line shows you the output from the validateCurrency function: it's returning a USD-flavored value from our Currency enumeration. Let's check that the case-insensitivity is working too:

```
scala> forex.CurrencyValidator1.validateCurrency("eur")
res1: forex.Currency.Value = EUR
```

That seems to be working. Now let's see what happens if we pass in an invalid value:

```
scala> forex.CurrencyValidator1.validateCurrency("dogecoin")
forex.UnsupportedCurrencyException: Currency must be USD/EUR/GBP, not dogecoin
  at forex.CurrencyValidator1$.validateCurrency(currency.scala:23)
  ... 28 elided
```

That looks correct. Trying to validate dogecoin as a currency is throwing an UnsupportedCurrencyException. We have ported our Java currency validator to Scala—but so far, we have only tweaked syntax; the semantics of our failure handling are the same. We can do better, as you'll see in the next iteration.

### 8.3.4 *Better failure handling through Scalaz*

Let's take a second pass at our currency validator. You can see the updated object, CurrencyValidator2, in the following listing. We haven't changed our Currency enumeration, and are no longer using the UnsupportedCurrencyException, so these are left out.

**Listing 8.7   CurrencyValidator2.scala**

```
package forex

import scalaz._          Imports the
import Scalaz._          Scalaz library

object CurrencyValidator2 {
                                              Our return type is a Scalaz
                                              Validation "box" around
  def validateCurrency(raw: String):          either a String or a Currency.
    Validation[String, Currency.Value] = {  ◁─┘

    val rawUpper = raw.toUpperCase(java.util.Locale.ENGLISH)    On success, we
    try {                                                       return our Currency
      Success(Currency.withName(rawUpper))                      inside a Success-
    } catch {                                               ◁── flavored Validation.
      case nsee: NoSuchElementException =>
        Failure("Currency must be USD/EUR/GBP and not " + raw)   ◁─┐
    }                                                              │
  }              On failure, we return our error String           │
}                inside a Failure-flavored Validation.            ┘
```

Again, here are a few notes to make the following listing a little easier to digest:

- This time, our file contains only a single object, so our filename matches the object.
- We no longer throw an exception so have no further need for the @throws annotation.
- Scala uses [] for specifying generics, whereas Java uses <>.

Most important, in place of throwing exceptions, we have moved our failure path into the return value of our function: this complex-looking Validation type, provided by Scalaz. Let's see how this Validation type behaves in the REPL; note that we have to quit and restart the console to force a recompile:

```
<Ctrl-D>
$ gradle repl --console plain
...
scala> forex.CurrencyValidator2.validateCurrency("eur")
res0: scalaz.Validation[String,forex.Currency.Value] = Success(EUR)

scala> forex.CurrencyValidator2.validateCurrency("dogecoin")
res1: scalaz.Validation[String,forex.Currency.Value] =
 Failure(Currency must be USD/EUR/GBP and not dogecoin)
```

You can think of Validation as a kind of box, or context, that describes whether the value inside it is on our happy path (called Success in Scalaz) or our failure path (called Failure). Even neater, the value inside the Validation box can have a different *type* on the Success side versus on the Failure side:

```
Validation[String, Currency.Value]
```

The first type given, String, is the type that we will use for our failure path; the second type, Currency.Value, is the type that we will use for our success path. Figure 8.9 shows how Validation is working, using the metaphor of cardboard boxes and paths.

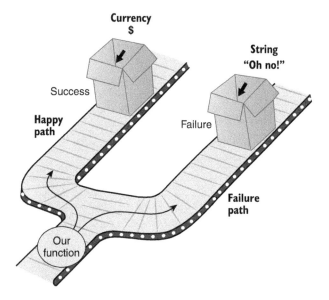

**Figure 8.9** For the happy path, our function returns a Currency boxed inside a Success. For the failure path, our function returns an error String boxed inside a Failure. Success and Failure are the two modes of the Scalaz Validation type.

This switch from using exceptions to using Scalaz's Validation to box either success or failure might not seem like a big change, but it's going to be our key building block for working with failure in Scala.

### 8.3.5 Composing failures

We're now happy with our function to validate that the incoming currency is one of our three supported currencies. But remember, another possible bug remains in our event stream:

> *Perhaps a hacker has managed to transact an order with a non-numeric order value, getting themselves a huge flat-screen TV for ¥l33t.*

We can put a stop to this with an AmountValidator object containing another validation function: one that checks that the incoming stringly typed order amount can be parsed to a Scala Double. This is set out in the following listing.

**Listing 8.8   AmountValidator.scala**

```
package forex

import scalaz._
import Scalaz._

object AmountValidator {

  def validateAmount(raw: String): Validation[String, Double] = {

    try {
      Success(raw.toDouble)
    } catch {
      case nfe: NumberFormatException =>
        Failure("Amount must be parseable to Double and not " + raw)
    }
  }
}
```

> Our return type is another Validation "box," around either a String or a Double.

> On success, we return our Double inside a Success.

> On failure, we return our error String inside a Failure.

By now, the pattern should be familiar: our function uses a Validation to represent our return value being either on the happy path (with a Success), or on the failure path (with a Failure). As before, we are returning different types of values inside our Success and our Failure: in this case, a Double or a String, respectively.

A quick check to make sure this is working in the Scala REPL:

```
<Ctrl-D>
$ gradle repl --console plain
...
scala> forex.AmountValidator.validateAmount("31.98")
res2: scalaz.Validation[String,Double] = Success(31.98)

scala> forex.AmountValidator.validateAmount("L33T")
res3: scalaz.Validation[String,Double] = Failure(Amount must be
 parseable to Double and not L33T)
```

Great! So we now have two validation functions:

- CurrencyValidator2.validateCurrency
- AmountValidator.validateCurrency

Both functions would have to return a Success in order for us to assemble a valid order total. If we get a Failure, or indeed two Failures, then we will find ourselves squarely on the failure path. Let's build a function now that attempts to construct an order total by running both of our validation functions. See the following listing for the code.

**Listing 8.9   OrderTotal.scala**

```
package forex

import scalaz._
import Scalaz._
```

```
case class OrderTotal(currency: Currency.Value, amount: Double)     ◁─┐
                                                           A case class representing
object OrderTotal {                                              our order total

  def parse(rawCurrency: String, rawAmount: String):
    Validation[NonEmptyList[String], OrderTotal] = {         Validates the
                                                             currency and
                                                             assigns to a val
    val c = CurrencyValidator2.validateCurrency(rawCurrency)  ◁─┘
    val a = AmountValidator.validateAmount(rawAmount)       ◁─┐
                                                             Validates the
    (c.toValidationNel |@| a.toValidationNel) {    ◁─┐      amount and assigns
      OrderTotal(_, )                                       to a val
    }                                  Combines our two
  }                               validations into the return
}                                   value for this function
```

Our parse method returns a Scalaz Validation.

A lot of new things are certainly happening in a small amount of code, but don't worry, we will go through this listing carefully in a short while. Before we do that, let's return to the Scala REPL and see how this `parse` function performs, first if both the currency and amount are valid:

```
<Ctrl-D>
$ gradle repl --console plain
...
scala> forex.OrderTotal.parse("eur", "31.98")
res0: scalaz.Validation[scalaz.NonEmptyList[String],forex.OrderTotal]
 = Success(OrderTotal(EUR,31.98))
```

This result makes sense: because both validations passed, we are staying on the happy path, and our `OrderTotal` is boxed in a `Success` to reflect this. Now let's see what happens if one or either of our validations fails:

```
scala> forex.OrderTotal.parse("dogecoin", "31.98")
res2: scalaz.Validation[scalaz.NonEmptyList[String],forex.OrderTotal]
 = Failure(NonEmpty[Currency must be USD/EUR/GBP and not dogecoin])

scala> forex.OrderTotal.parse("eur", "L33T")
res3: scalaz.Validation[scalaz.NonEmptyList[String],forex.OrderTotal]
 = Failure(NonEmpty[Amount must be parseable to Double and not L33T])
```

In both cases, we end up with a `Failure` containing a `NonEmptyList`, which in turn contains our error message as a `String`. In fact, `NonEmptyList` is another Scalaz type. It's similar to a standard Java or Scala `List`, except that a `NonEmptyList` (sometimes called a Nel, or NEL, for short) cannot be, well, empty. This suits our purposes well; if we have a `Failure`, we know at least one cause of it, and so our list of error messages will never be empty.

Why do we need a NEL on the `Failure` side in the first place? Hopefully, this next test should make the reason clear:

```
scala> forex.OrderTotal.parse("dogecoin", "L33T")
res4: scalaz.Validation[scalaz.NonEmptyList[String],forex.OrderTotal]
```

```
= Failure(NonEmpty[Currency must be USD/EUR/GBP and not dogecoin,
Amount must be parseable to Double and not L33T])
```

It's a little bit long-winded, but you can see that the value returned from the parse function is now a Failure dutifully recording both error messages in String form; this approach is similar to the way you might see multiple validation errors (Phone Is Missing Country Code, and so forth) on a website form when you click the Submit button.

Now that you understand what the function does, let's return to the code itself:

```
(c.toValidationNel |@| a.toValidationNel) {
  OrderTotal(_, _)
}
```

What exactly is going on here, and what is that mysterious |@| operator doing? The first thing to explain is the toValidationNel method calls. There is not much mystery here: this is a method available to any Scalaz Validation that will promote the Failure value into a NEL. Here's a quick demonstration in the Scala REPL:

```
scala> import scalaz._
import scalaz._

scala> import Scalaz._
import Scalaz._

scala> val failure = Failure("OH NO!")
failure: scalaz.Failure[String] = Failure(OH NO!)

scala> failure.toValidationNel
res8: scalaz.ValidationNel[String,Nothing] = Failure(NonEmpty[OH NO!])
```

See how the error message is now inside a NEL? This happens only in the case of Failure. If we have a Success, the value inside is untouched, although the overall type of the Validation will still change:

```
scala> val success = Success("WIN!")
success: scalaz.Success[String] = Success(WIN!)

scala> success.toValidationNel
res6: scalaz.ValidationNel[Nothing,String] = Success(WIN!)
```

Now onto the |@| operator. Unfortunately, there is no official Scalaz documentation about this, but if you dig into the code, you will find it referred to as follows:

*[a] DSL for constructing Applicative expressions*

The |@| operator is sometimes referred to as the *Home Alone* or *chelsea bun* operator, but we prefer to call it the *Scream* operator after the Munch painting. Regardless, this operator is doing something clever: it is composing our two input Validations into a new output Validation, intelligently determining whether the output Validation should be a Success or a Failure based on the inputs. Figure 8.10 shows the different options.

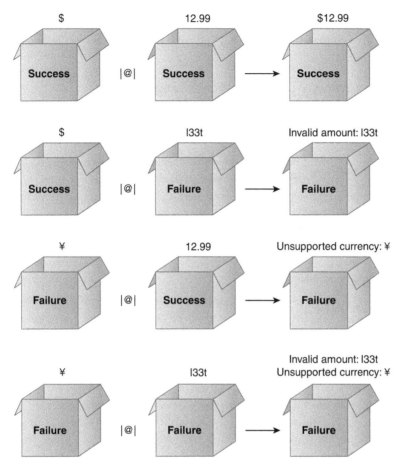

**Figure 8.10** **Applying the Scream operator to two Scalaz** Validations **gives us a matrix of possible outputs, depending on whether each input** Validation **is a** Success **or** Failure. **Note that we get an output of** Success **only if both inputs are** Success.

As you can see in figure 8.10, the Scream operator allows us to compose multiple Validation inputs into a single Validation output. The example uses two inputs just to keep things simple: we could just as easily compose nine or twenty Validation inputs into a single Validation output. If we composed twenty Validation inputs with the Scream operator, *all* of our twenty inputs must be a Success for us to end up with a Success output. On the other hand, if all of our twenty inputs were Failures containing a single error String, then our output would be a Failure containing a NonEmptyList of twenty error Strings.

You have seen how to handle failure inside a single processing step: we can perform multiple pieces of validation in parallel, and then compose the successes or failures into a single output. If this is failure processing *in the small*, how do we start to handle failures between different processing steps inside the same job? We'll look at this next. Feel free to grab a cup of coffee before you apply what you've learned so far on our unified log.

## 8.4    *Implementing railway-oriented processing*

> *Everything has an end, and you get to it if you only keep all on.*
>
> —E. Nesbit, *The Railway Children*

A common theme running through this chapter has been the idea of our code embarking on a happy path and dropping down to a failure path if something (or multiple things) goes wrong. We have looked at this pattern at a large scale, across multiple stream processing jobs, as well as at a small scale, at the level of composing multiple failures inside a single processing step. Now let's look at the middle ground: handling failure across multiple processing steps inside a single job.

### 8.4.1    *Introducing railway-oriented processing*

In the preceding section, the Scream operator let us compose an output from separate pieces of processing into a single unified output. Specifically, we composed the Validation outputs from our currency and amount parsing into a single Validation output. Where do we go from here?

Remembering back to our original brief, we also need the current exchange rate between our order's currency and our employer's base currency. Again, this requirement could also take us onto the failure path: fetching the exchange rate could fail for various reasons. We have several processing steps now, so let's sketch out an end-to-end happy path that incorporates all of this processing. Figure 8.11 presents two separate versions of this happy path:

- The *idealized happy path*—This expresses the dependencies between each piece of work. Specifically, we need a successfully validated currency to look up the currency, but we don't need the validated amount until we are trying to perform the conversion.
- The *pragmatic happy path*—We look up the currency only if both the currency and the amount validate successfully.

The pragmatic happy path is so-called because looking up a currency from a third-party service's API over HTTP is an *expensive* operation, whereas validating our order amount is relatively cheap. We could be performing this processing job for many millions of events, so even small unnecessary delays will add up; therefore, if we can fail fast and save ourselves some pointless currency lookups, we should.

Figure 8.11 shows us only the end-to-end happy path, resulting in a customer order successfully converted into the retailer's base currency. Figure 8.12 provides the next

**Idealized graph**

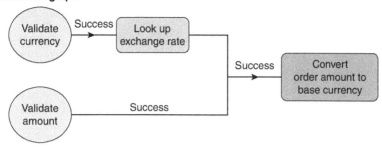

**Pragmatic graph**

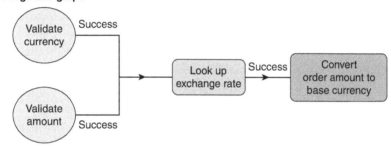

**Figure 8.11** The idealized and pragmatic happy paths vary based on how late the exchange rate lookup is performed. In the pragmatic happy path, we validate as much as we can before attempting the exchange rate lookup.

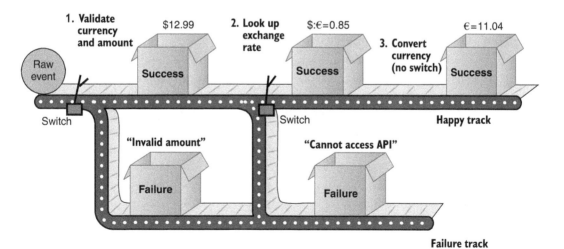

**Figure 8.12** Processing of our raw event proceeds down the happy track, but a failure at any given step will switch us onto the failure track. When we are on the failure track, we stay there, bypassing any further happy-track processing.

level of detail, imagining our processing job as a railway line (or conveyor belt), with two tracks:

- *The happy track*—This shows the type transformations that occur as we successfully validate first the currency and amount (one processing step), then successfully look up the exchange rate (our second processing step), and finally convert the order amount.
- *The failure track*—We can be switched onto this track if any of our processing steps fail.

This is not my own metaphor: *railway-oriented programming* was coined by functional programmer Scott Wlaschin in his eponymous blog post (https://fsharpforfunand-profit.com/posts/recipe-part2/). Scott's blog post uses the railway metaphor with happy and failure paths to introduce compositional failure handling in the F# programming language. Another functional language, F# enables a similar approach to failure processing to our Scala-plus-Scalaz combination. We strongly recommend reading Scott's post after you have finished this chapter; in the meantime, here is railway-oriented programming in Scott's own words:

> *What we want to do is connect the* Success *output of one to the input of the next, but somehow bypass the second function in case of a* Failure *output....There is a great analogy for doing this—something you are probably already familiar with. Railways! Railways have switches ("points" in the UK) for directing trains onto a different track. We can think of these "Success/Failure" functions as railway switches....We will have a series of black-box functions that appear to be straddling a two-track railway, each function processing data and passing it down the track to the next function....Note that once we get on the failure path, we never (normally) get back onto the happy path. We just bypass the rest of the functions until we reach the end.*

The railway-oriented approach has important properties that might not be immediately obvious but affect how we use it:

- It *composes* failures within an individual processing step, but it *fails fast* across multiple steps. This makes sense: if we have an invalid currency code, we can validate the order amount as well (same step), but we can't proceed to getting an exchange rate for the corrupted code (the next step).
- The type inside our Success can change between each processing step. In figure 8.12, our Success box contains first an OrderTotal, then an exchange rate, and then finally an OrderTotal again.
- The type inside our Failure must stay the same between each processing step. Therefore, we have to choose a type to record our failures and stick to it. We have used a NonEmptyList[String] in this chapter because it is simple and flexible.

Enough of the theory for now—let's build our railway in Scala.

### 8.4.2 Building the railway

First, we need a function representing our exchange-rate lookup, which could fail. Because our focus in this chapter is on failure handling, our function will contain some hardcoded exchange rates plus some hardwired "random failure." If you are interested in looking up real exchange rates in Open Exchange Rates, you can always check out Snowplow's scala-forex project (https://github.com/snowplow/scala-forex/).

The code for our new function, `lookup`, is in the following listing. The function consists of a single pattern-match expression that returns the following:

- On the happy track, an exchange rate `Double` boxed in a `Success`
- On the failure track, an error `String` boxed in a `Failure`

> **Listing 8.10  ExchangeRateLookup.scala**

```
package forex

import util.Random

import scalaz._
import Scalaz._

object ExchangeRateLookup {

  def lookup(currency: Currency.Value):
    Validation[String, Double] = {            ◁──  Returns either a Failure
                                                   String or a Success Double
                                                   (the exchange rate)

    currency match {
      case Currency.Eur       => 1D.success           .success and .failure
      case _ if isUnlucky() => "Network error".failure   are Scalaz "sugar" for
      case Currency.Usd       => 0.85D.success         Success() and Failure().
      case Currency.Gbp       => 1.29D.success
      case _                  => "Unsupported currency".failure
    }
  }

  private def isUnlucky(): Boolean =
    Random.nextInt(5) == 4
}
```

**Should never happen, but we want an exhaustive pattern match** (points to `case _`)

**Simulates a network error when retrieving USD or GBP exchange rate**

If you haven't seen a pattern match before, you can think of it as similar to a C- or Java-like switch statement, but much more powerful. We can match against specific cases such as `Currency.Eur`, and we can also perform wildcard matches using `_`. The `if isUnlucky()` is called a *guard*; it makes sure that we match this pattern only if the condition is met. As you can see from the definition of the `isUnlucky` function, we are simulating a network failure that should happen 20% of the time when we are looking up the exchange rate over the network.

An important point on the final `_` wildcard match: we include this to ensure that our pattern match is *exhaustive*—that all possible patterns that could occur are handled.

Without this final wildcard match, the code would still compile, but we are sitting on a potential problem if the following occur:

1  A new currency (for example, Yen) is added to the Currency enumeration.
2  Support for the new currency is added to the validateCurrency function.
3  But we forget to add this new currency into this pattern-match statement.

In this case, the pattern match would throw a runtime MatchError on every yen lookup, causing our stream processing job to crash. We want to avoid this, so our final _ wildcard match defends against this *pre-bug*, a bug that has not happened yet.

Let's put our new function through its paces in the Scala REPL:

```
<Ctrl-D>
$ gradle repl --console plain
...
scala> import forex.{ExchangeRateLookup, Currency}
import forex.{ExchangeRateLookup, Currency}

scala> ExchangeRateLookup.lookup(Currency.Usd)
res2: scalaz.Validation[String,Double] = Success(0.85)

scala> ExchangeRateLookup.lookup(Currency.Gbp)
res2: scalaz.Validation[String,Double] = Success(1.29)
```

Great—that's all working fine. Our network connection is also suitably unreliable; every so often when looking up a rate, you should see an error like this:

```
scala> ExchangeRateLookup.lookup(Currency.Gbp)
res1: scalaz.Validation[String,Double] = Failure(Network error)
```

Now, remembering back to section 8.3, we have two functions that represent the two distinct steps in our processing, either of which could fail. Here are the signatures of both functions:

- OrderTotal.parse(rawCurrency: String, rawAmount: String): Validation[NonEmptyList[String], OrderTotal]
- ExchangeRateLookup.lookup(currency: Currency.Value): Validation[String, Double]

Note that the types on our failure track are not quite the same: a NonEmptyList of Strings, versus a singular String. In theory, this sounds like it breaks the rule that "every Failure must contain the same type," but in practice it is easy to promote a single String to be a NonEmptyList of Strings.

Remember that we want to try to generate an order total first, and only then look up our exchange rate. Let's create a new function, OrderTotalConverter1.convert, which captures all of the processing we need to do in this job. The signature of this function should look like this:

```
convert(rawCurrency: String, rawAmount: String): ValidationNel[String,
    OrderTotal]
```

You haven't seen ValidationNel[A, B] before. This is Scalaz shorthand for Validation[NonEmptyList[A], B]. Putting this all together: we will attempt to generate an
OrderTotal in our base currency from an incoming raw currency and order total. If
we somehow end up on the failure path, we will return a NonEmptyList of the error
Strings that we encountered. Let's see an implementation of this function in the following listing.

**Listing 8.11   OrderTotalConverter1.scala**

```
package forex

import scalaz._
import Scalaz._                              Aliases to make our
                                             function more readable
import forex.{OrderTotal => OT}        ◁┘
import forex.{ExchangeRateLookup => ERL}

object OrderTotalConverter1 {

  def convert(rawCurrency: String, rawAmount: String):
    ValidationNel[String, OrderTotal] = {           ◁─┐  Parse our raw currency
                                                      │  and amount into a
    for {                                             │  Validation.
      total <- OT.parse(rawCurrency, rawAmount)     ◁─┘
      rate  <- ERL.lookup(total.currency).toValidationNel  ◁─ Look up our
      base   = OT(Currency.Eur, total.amount * rate) ◁─┐        exchange rate into
    } yield base                                       │        a Validation.
  }                          Calculate our order total in
}                               the base currency.
```

There is some interesting new syntax here; that for {} yield is clearly not your grand-
father's for loop! Before we dive into this, let's fire up the Scala REPL one last time
for this chapter and put this function through its paces. First let's stick to the happy
path (assuming our unreliable network lets us):

```
<Ctrl-D>
$ gradle repl --console plain
...
scala> forex.OrderTotalConverter1.convert("usd", "12.99")
res1: scalaz.ValidationNel[String,forex.OrderTotal]
 = Success(OrderTotal(EUR,11.0415))

scala> forex.OrderTotalConverter1.convert("EUR", "28.98")
res2: scalaz.ValidationNel[String,forex.OrderTotal]
 = Success(OrderTotal(EUR,28.98))
```

Now let's see about the failure path:

```
scala> forex.OrderTotalConverter1.convert("yen", "l33t")
res3: scalaz.ValidationNel[String,forex.OrderTotal]
 = Failure(NonEmptyList(Currency must be USD/EUR/GBP and not yen,
 Amount must be parseable to Double and not l33t))
```

```
scala> forex.OrderTotalConverter1.convert("gbp", "49.99")
res56: scalaz.ValidationNel[String,forex.OrderTotal]
 = Failure(NonEmptyList(Network error))
```

You might have to repeat the second conversion a few times to see the Network Error `Failure` caused by the exchange rate lookup. Also, remember that you will see the network error only if the raw currency and amount are valid: if either is invalid, we have already failed fast, and the exchange rate will not be looked up.

So our `convert()` function is working, but how is it working? The key is to understand the `for {} yield` construct. Wherever you see the `for` keyword in Scala, you are looking at a Scala `for` comprehension, which is syntactic sugar over a set of Scala's functional operations: `foreach`, `map`, `flatMap`, `filter`, or `withFilter`.[3] This is modeled on the `do` notation found in Haskell, a pure functional language.

Scala will translate our `for {} yield` into a set of `flatMap` and `map` operations, as shown in listing 8.12. The code in `OrderTotalConverter2` is functionally identical to our previous `OrderTotalConverter1`; you can test it at the Scala REPL if you like:

```
<Ctrl-D>
$ gradle repl --console plain
...
scala> forex.OrderTotalConverter2.convert("yen", "l33t")
res3: scalaz.ValidationNel[String,forex.OrderTotal]
 = Failure(NonEmpty[Currency must be USD/EUR/GBP and not yen,
 Amount must be parseable to Double and not l33t])
```

**Listing 8.12   OrderTotalConverter2.scala**

```
package forex

import scalaz._
import Scalaz._
import scalaz.Validation.FlatMap._

import forex.{OrderTotal => OT}
import forex.{ExchangeRateLookup => ERL}

object OrderTotalConverter2 {

  def convert(rawCurrency: String, rawAmount: String):       We replace our first
    ValidationNel[String, OrderTotal] = {                    <- with a .flatMap.

    OT.parse(rawCurrency, rawAmount).flatMap(total =>     <──┐
      ERL.lookup(total.currency).toValidationNel.map((rate: Double) =>  <──┐
        OT(Currency.Eur, total.amount * rate)))    <──┐
  }                                                     No need to assign a      We replace our second
}                                                       name like "base" here   <- with a .map.
```

---

[3]  More information on how the Scala yield keyword works can be found at https://docs.scala-lang.org/tutorials/FAQ/yield.html.

Putting them side by side, OrderTotalConverter1 is much more readable than OrderTotalConverter2, but this second version gives us a better starting point for understanding how we have implemented this fail-fast approach across multiple processing steps. This is the last piece of the railway-oriented processing puzzle.

Also note that OrderTotalConverter1 does work only when compiled against Scala 2.11, as Scala 2.12 has introduced enhanced type checking that made type inference inside for comprehensions more difficult to achieve.

flatMap and map are the last two pieces in our toolkit for this chapter, so we'll go through these in some detail. We'll start with map as it's simpler to understand. Here is a simplified definition for the map method on a Validation[F, S] called self:

```
def map(aFunc: S => T): Validation[F, T] = self match {
  case Success(aValue) => Success(aFunc(aValue))
  case Failure(fValue) => Failure(fValue)
}
```

Because this is Scala, we are able to define map in terms of a single pattern-match expression. The way to read aFunc: S => T is as a function that takes an argument of type S and returns a value of type T. If our Validation is a Success, we apply the function supplied to map to the value contained inside the Success, resulting in a new value, possibly of a new type, but still safely wrapped in our Validation container. To put it another way: if we start with Success[S] and apply a function S => T to the value inside Success, we end up with Success[T]. On the other hand, if our Validation is a Failure, we leave this as is: a Failure[F] stays a Failure[F], containing the exact same value.

Mapping over a Validation is visualized in figure 8.13, for both a Success and a Failure.

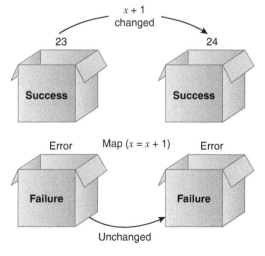

**Figure 8.13** If we map a simple function (adding 1 to a number) to a Success-boxed 23 and a Failure-boxed error String, the Failure box is untouched, but the Success box now contains 24.

Now let's take a look at `flatMap`. Like `map`, `flatMap` takes a function, applies it to a `Validation[F, S]`, and returns a `Validation[F, T]`. The difference is in the function that `flatMap` takes as its argument: the function has the type `S => Validation[F, T]`. In other words, `flatMap` expects a function that takes a value and produces a new value inside a new `Validation` container. You may be thinking, hang on a moment—why does a `flatMap`, like a `map`, return this

```
Validation[F, T]
```

and not this:

```
Validation[F, Validation[F, T]], given the supplied function
```

The answer to this question lies in the *flat* in the name: `flatMap` *flattens* the two `Validation` containers into one. `flatMap` is the secret sauce of our railway-oriented approach: it allows us to chain multiple processing steps together into one stream processing job without accumulating another layer of `Validation` boxing at every stage. Figure 8.14 depicts the unworkable alternative.

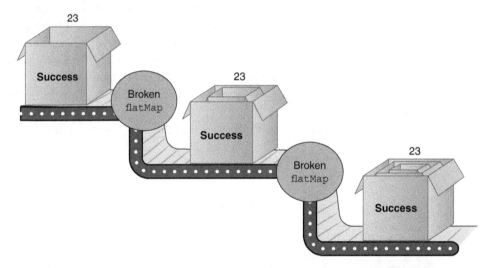

**Figure 8.14  If our `flatMap` did not flatten, we would end up with a Matryoshka-doll-like collection of `Validation`s inside other `Validation`s. It would be extremely difficult to work with this nested type correctly.**

The simplified definition for `flatMap` on a `Validation[F, S]` called `self` should look familiar after `map`:

```
def flatMap(aFunc: S => Validation[F, T]): Validation[F, T] = self match {
  case Success(aValue) => aFunc(aValue)
  case Failure(fValue) => Failure(fValue)
}
```

The two differences from map are as follows:

- The function passed to flatMap returns a Validation[F, T], not just a T.
- On Success, we remove the original Success container and leave it to aFunc to determine whether T ends up in a new Success or Failure.

Putting this all together, figure 8.15 attempts to visualize how our convert function is working.

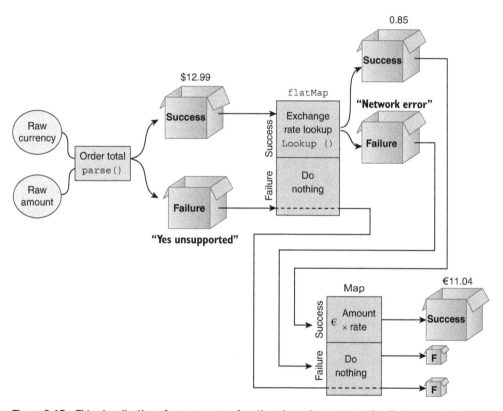

**Figure 8.15** **This visualization of our** convert **function shows how our now-familiar** Validation **boxes interact first with a** flatMap **and then with a** map. flatMap **allows us to chain together multiple steps that each might result in** Failure, **without nesting** Validations **inside other** Validations. **Both** flatMap **and** map **are respecting the fail-fast requirement of the pipeline, with no further processing after a Failure has occurred.**

The key point to understand here is that flatMap and map support our railway-oriented processing approach across multiple work steps:

- We can compose Validations that are peers of each other *inside* an individual work step by using the Scream operator introduced in section 8.3.
- We can then compose multiple sequential steps that depend on the previous steps succeeding by using flatMap and map as required.

- If we are lucky enough to stay on the happy path, we can transform the value and type, safely boxed in our Success, as we move from work step to work step.
- Conversely, as soon as we encounter a Failure at the boundary between one work step and the next, we can fail fast, with our Failure-boxed error or errors becoming the final result of our processing flow.

This completes our look at failure handling in the unified log. If we take a moment's breath and look back over the previous three sections, we see an interesting pattern emerging, one that you could call the *compose, fail-fast, compose* sandwich (or burger, if you prefer). To explain myself:

- *Composition in the large*—We compose complex event-stream-processing work-flows out of multiple stream-processing jobs, each of which outputs a happy stream and a failure stream.
- Fail-*fast as the filling or patty*—If a stream processing job consists of multiple work steps that have a dependency chain between them, we must fail fast as soon as we encounter a step that did not succeed. Scala's flatMap and map help us to do this.
- *Composition in the small*—If we have a work step inside our job that contains mul-tiple independent tasks, we can perform all of these and then compose these into a final verdict on whether the step was a Success or a Failure. Scalaz's Scream operator, |@|, helps us to do this.

## Summary

- Unix's concept of three standard streams, one for input and one each for good and bad output, is a powerful idea that we can apply to our own unified log processing.
- Java uses exceptions for program termination or recovery but lacks tools to ele-gantly address failure at the level of an individual *unit of work*. Many program-mers lean on error logging to address this, which has helped to spawn the *log-industrial complex*.
- In the unified log, we can model failures as events themselves. These events should contain all of the causes of the failure, and should also contain the orig-inal event, to enable reprocessing in another stream-processing job.
- Stream processing jobs should echo the Unix approach and write successes to one stream, and failures to another stream. This allows us to compose complex processing flows out of multiple jobs.
- As a strongly typed functional language, Scala provides tools such as flatMap, map, and for {} yield syntactic sugar, which help us to keep our failure path integrated into our core code flow, versus exception-throwing or error-logging approaches.
- The Scalaz Validation is a kind of container that can represent either Success or Failure and can contain different types for each. Again, this helps us to keep our failure path co-situated with our happy path.

- The Scream operator, `|@|`, can be applied to multiple Scalaz `Validations` to compose an output `Validation`. This lets us compose multiple tasks inside a single step, into a final verdict on whether that step succeeded or failed.
- Across multiple work steps inside a single job, we want to fail fast as soon as we encounter a step that did not succeed. Scala's `flatMap` and `map` work with the `Validation` type to let us do this; Scala's `for {} yield` syntax makes our code much more readable.
- If we lean back and squint a little, we can see the overall methodology as a compose, fail-fast, compose approach, where we compose multiple jobs, fail fast between multiple steps inside the same job, and again compose between multiple tasks within the same job step.

# Commands

*9*

---

**This chapter covers**

- Understanding the role of commands in the unified log
- Modeling commands
- Using Apache Avro to define schemas for commands
- Processing commands in our unified log

So far, we have concerned ourselves almost exclusively with events. Events, as we explained in chapter 1, are discrete occurrences that take place at a specific point in time. Throughout this book, we have created events, written them to our unified log, read them from our unified log, validated them, enriched them, and more. There is little that we have not done with events!

But we can represent another unit of work in our unified log: the *command*. A command is an order or instruction for a specific action to be performed in the future—each command, when executed, produces an event. In this chapter, we will show how representing commands explicitly as one or more streams within our unified log is a powerful processing pattern.

We will start with a simple definition for commands, before moving on to show how decision-making apps can operate on events in our unified log and express decisions in the form of commands. Modeling commands for maximum flexibility and utility is something of an art; we will cover this next before launching into the chapter's applied example.

Working for Plum, our fictitious global consumer-electronics manufacturer, we will define a command intended to alert Plum maintenance engineers of overheating machines on the factory floor. We will represent this command in the Apache Avro schema language and use the Kafka producer script from chapter 2 to publish those alert commands to our unified log.

We will then write a simple command-executor app by using the Kafka Java client's producer and consumer capabilities. Our command executor will read the Kafka topic containing our alert commands and send an email to the appropriate support engineer for each alert; we will use Rackspace's Mailgun transactional email service to send the emails.

Finally, we will wrap up the chapter with some design questions to consider before adding commands into your company's own unified log. We will consider the pros and cons of two designs for your command topic(s) in Kafka or Kinesis and will also introduce the idea of a hierarchy of commands.

Let's get started!

## 9.1 Commands and the unified log

What is a command exactly, and what does it mean in the context of the unified log? In this section, we'll introduce commands as complementary to events, and argue for making decision-making apps, and the commands they generate, a central part of your unified log.

### 9.1.1 Events and commands

A *command* is an order or instruction for a specific action to be performed in the future. Here are some example commands:

- "Order a pizza, Dad."
- "Tell my boss that I quit."
- "Renew our household insurance, Jackie."

If an event is a record of an occurrence in *the past*, then a command is the expression of intent that a new event will occur *in the future*:

- My dad ordered a pizza.
- I quit my job.
- Jackie renewed our household insurance.

Figure 9.1 illustrates the behavioral flow underpinning the first example: a *decision* (I want pizza) produces a *command* ("Order a pizza, Dad"), and if or when that command is *executed*, then we can record an event as having taken place (my dad ordered a pizza). In grammatical terms: an event is a *past-tense* verb in the *indicative mood*, whereas a command is a verb in the *imperative mood*.

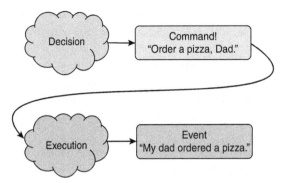

**Figure 9.1  A decision produces a command—an order or instruction to do something specific. If that command is then executed, we can record an event as having occurred.**

You can see that there is a symbiotic relationship between commands and events: strictly speaking, without commands, we would have no events.

### 9.1.2   *Implicit vs. explicit commands*

If behind every event is a command, how could we have reached chapter 9 in a book all about events without encountering a single command? The answer is that almost all software relies on *implicit commands*—the decision to do something is immediately followed by the execution of that decision, all in the same block of code. Here is a simple example of this in pseudocode:

```
function decide_and_act(obj):
    if obj is an apple:
        eat the apple
        emit_event(we ate the apple)
    else if obj is a song:
        sing the song
        emit_event(we sang the song)
    else if obj is a beer:
        drink the beer
        emit_event(we drank the beer)
    else:
        emit_event(we don't recognize the obj)
```

This code includes no explicit commands: instead, we have a block of code that mixes decision-making in with the execution of those decisions. How would this code look if we had explicit commands? Something like this:

```
function make_decision(obj):
    if obj is an apple:
        return command(eat the apple)
    else if obj is a song:
        return command(sing the song)
    else if obj is a beer:
        return command(drink the beer)
    else:
        return command(complain we don't recognize the obj)
```

Of course, this version is not functionally identical to the previous code block. In this version, we are only *returning commands* as the result of our decisions; we are not actually executing those commands. There will have to be another piece of code downstream of make_decision(obj), looking something like this:

```
function execute_command(cmd):
    if cmd is eat the apple:
        eat the apple
        emit_event(we ate the apple)
    else if cmd is sing the song:
        sing the song
        emit_event(we sang the song)
    else if cmd is drink the beer:
        drink the beer
        emit_event(we drank the beer)
    else if cmd is complain we don't recognize the obj
        emit_event(we don't recognize the obj)
```

It looks like we have turned something simple—making a decision and acting on it—into something more complicated: making a decision, emitting a command, and then executing that command. The reason we do this is in support of *separation of concerns.*[1] Turning our commands into explicit, first-class entities brings major advantages:

- *It makes our decision-making code simpler.* All make_decision(obj) needs to be able to do is emit a command. It doesn't need to understand the mechanics of eating apples or singing songs; nor does it need to know how to track events.
- *It makes our decision-making code easier to test.* We have to test only that our code makes the right decision, not that it then executes on that decision correctly.
- *It makes our decision-making process more auditable.* All decisions lead to concrete commands that can be reviewed. By contrast, with decide_and_act(obj), we have to explore the *outcomes* of the actions to understand what decisions were made.
- *It makes our execution code more DRY ("don't repeat yourself").* We can implement eating an apple in a single code location, and any code that decides to eat an apple emits a command instructing the eating.
- *It makes our execution code more flexible.* If our command is "send Jenny an email," we can swap out one transactional email service provider (for example, Mandrill) for another (for example, Mailgun or SendGrid).
- *It makes our execution code repeatable.* We can replay a sequence of commands in the same order as they were issued at any time if the context requires it.

This decoupling of decision-making and command execution clearly has benefits, but if we involve our unified log, things get even more powerful. Let's explore this next.

---

[1] You can learn more about separation of concerns at Wikipedia: https://en.wikipedia.org/wiki/Separation_of_concerns.

### 9.1.3   *Working with commands in a unified log*

In the previous section, we split a single function, decide_and_act(obj), into two separate functions:

- make_decision(obj)—Made a decision based on the supplied obj and returned a command to execute
- execute_command(cmd)—Executed the supplied command cmd and emitted an event recording the execution

With a unified log, there is no reason that these two functions need to live in the same application: we can create a new stream called *commands* and write the output of make_decision(obj) into that stream. We can then write a separate stream-processing job that consumes the *commands* stream and executes the commands found within by using execute_command(cmd). This is visualized in figure 9.2.

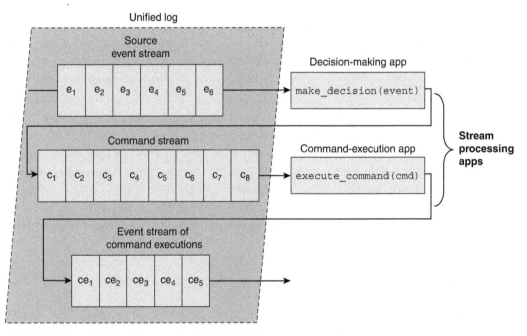

**Figure 9.2   Within our unified log, a source event stream is used to drive a decision-making app, which emits commands to a second stream. A command-execution app reads that stream of commands, executes the commands, and then emits a stream of events recording the command executions.**

With this architecture, we have completely decoupled the decision-making process from the execution of the commands. Our stream of commands is now a *first-class entity*: we can attach whatever applications we want to execute (or even just observe) those commands.

In the next section, we will add commands to our unified log.

## 9.2 *Making decisions*

To generate commands, we'll first have to make some decisions. In this section, we'll continue working with Plum, our fictional company with a unified log introduced in chapter 7, and wire some decision-making into that unified log. Let's get started.

### 9.2.1 *Introducing commands at Plum*

Let's imagine again that we are in the BI team at Plum, a global consumer-electronics manufacturer. Plum is in the process of implementing a unified log; for unimportant reasons, the unified log is a hybrid, using Amazon Kinesis for certain streams and Apache Kafka elsewhere. In reality, this is a fairly unusual setup, but it enables us to work with both Kinesis and Kafka in this section of the book.

At the heart of the Plum production line is the NCX-10 machine, which stamps new laptops out of a single block of steel. Plum has 1,000 of these machines in each of its 10 factories. Each machine emits key metrics as a form of health check to a Kafka topic every 5 minutes, as shown in figure 9.3.

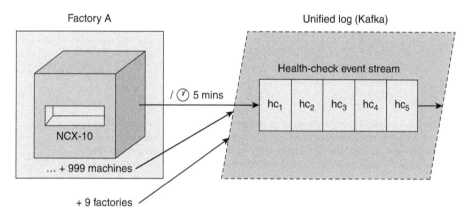

**Figure 9.3  All of the NCX-10 machines in Plum's factories emit a standard health-check event to a Kafka topic every 5 minutes.**

We have been asked by the plant maintenance team at Plum to use the health-check data to help keep the NCX-10 machines humming. The team members are particularly concerned about one scenario that they want to tackle first: if a machine shows signs of overheating, a maintenance engineer needs to be alerted immediately in order to be able to inspect the machine and fix any issues with the cooling system.

We can deliver what plant maintenance wants by writing two stream-processing jobs:

- *The decision-making job*—This will parse the event stream of NCX-10 health checks to detect signs of machines overheating. If these are detected, this job will emit a command to alert a maintenance engineer.

- *The command-execution job*—This will read the event stream containing commands to alert a maintenance engineer, and will send each of these alerts to the engineer.

Figure 9.4 depicts these two jobs and the Kafka streams between them.

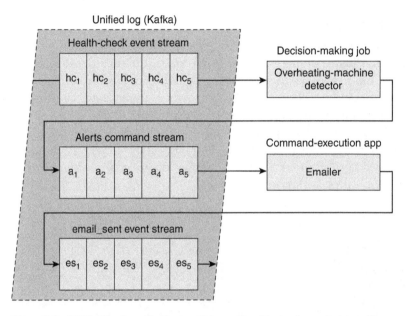

**Figure 9.4   Within Plum's unified log, a stream of health-check events is read by a decision-making app to detect overheating machines and emit alert commands to a second stream. A command-execution app reads the stream of commands and sends emails to alert the maintenance engineers.**

Does the decision-making job sound familiar? It's likely a piece of either single or multiple (stateful) event processing, as we explored in part 1. Because this chapter is about commands, we'll skip over the implementation of the decision-making job and head straight to defining the command itself.

### 9.2.2   *Modeling commands*

Our decision-making job needs to emit a command to alert a maintenance engineer that a specific NCX-10 machine shows signs of overheating. It's important that we model this command correctly: it will represent the *contract* between the decision-making job, which will emit the command, and the command-executing job, which will read the command.

We know what this command needs to do—alert a maintenance engineer—but how do we decide what fields to put in our command? Like a tasty muffin, a good command should have these characteristics:

- *Shrink-wrapped*—The command must contain everything that could possibly be required to execute the command; the executor should not need to look up additional information to execute the command.
- *Fully baked*—The command should define exactly the action for the executor to perform; we should not have to add business logic into the executor to turn each command into something actionable.

Here is an example of a badly designed self-describing alert command for Plum:

```
{ "command": "alert_overheating_machine",
  "recipient": "Suzie Smith",
  "machine": "cz1-123",
  "createdAt": 1539576669992
}
```

Let's consider how the executor would have to be written for this command, in pseudocode:

```
function execute_command(cmd):
    if cmd.type == "alert_overheating_machine":
        email_address = lookup_email_address(cmd.recipient)
        subject = "Overheating machine"
        message = "Machine {{cmd.machine}} may be overheating!"
        send_email(email_address, subject, message)
        emit_event(alerted_the_maintenance_engineer)
```

Unfortunately, this breaks both of the rules for a good command:

- Our command executor has to look up a crucial piece of information, the engineer's email address, in another system.
- Our command executor has to include business logic to translate the command type plus the machine ID into an actionable alert.

Here is a better version of an alert command for Plum:

```
{ "command" : "alert",
    "notification": {
    "summary": "Overheating machine",
    "detail": "Machine cz1-123 may be overheating!",
    "urgency": "MEDIUM"
  },
  "recipient": {
    "name": "Suzie Smith",
    "phone": "(541) 754-3010",
    "email": "s.smith@plum.com"
  },
  "createdAt": 1539576669992
}
```

The command may be a little more verbose, but the executor implementation for this command is much simpler:

```
function execute_command(cmd):
    if cmd.command == "alert":
```

```
send_email(cmd.recipient.email, cmd.notification.summary,
    cmd.notification.detail)
emit_event(alerted_the_maintenance_engineer)
```

This command requires the executor to do much less work: the executor now only has to send the email to the recipient and record an event to the same effect. The executor does not need to know that this specific alert relates to an overheating machine in a factory; the specifics of this alert were "fully baked" upstream, in the business logic of the decision-making job. This strong decoupling has the major advantage of making our executor *general-purpose.* Plum can write other decision-making jobs that can reuse this command-executing job without requiring any code changes in the executor.

The other thing to note is that we have made the alert command *definitive* but not overly *restrictive.* If we wanted to, we could update our executor to include a prioritization system for processing alerts, like so:

```
function execute_command(cmd):
    if cmd.command == "alert":
        if cmd.urgency == "HIGH":
            send_sms(cmd.recipient.phone, cmd.notification.detail)
        else:
            send_email(cmd.recipient.email, cmd.notification.summary,
                cmd.notification.detail)
            emit_event(alerted_the_maintenance_engineer)
```

Using our unified log for commands makes updates like this achievable:

- We only have to update our command-executor job to add the prioritization logic. There's no need to update the decision-making jobs in any way.
- We can test the new functionality by replaying old commands through the new executor.
- We can switch over to the new executor gradually, potentially running it in parallel to the old executor and allocating only a portion of commands to the new executor.

### 9.2.3  *Writing our alert schema*

We are now ready to define a *schema* for our alert command. We are going to use Apache Avro (https://avro.apache.org) to model this schema. As explained in chapter 6, Avro schemas can be defined in plain JSON files. The schema file needs to live somewhere, so we'll add it into our command-executor app. We'll need to create this app next.

We are going to write the command executor as a Java application, called Executor-App, using Gradle. Let's get started. First, create a directory called plum. Then switch to that directory and run this:

```
$ gradle init --type java-library
...
BUILD SUCCESSFUL
...
```

Gradle has created a skeleton project in that directory, containing a couple of Java source files for stub classes, called Library.java and LibraryTest.java. Delete these two files, as we will be writing our own code shortly.

Next let's prepare our Gradle project build file. Edit the file build.gradle and replace its current contents with the following listing.

**Listing 9.1   build.gradle**

```
plugins {                                      A Gradle plugin to generate Java
    id "java"                                  code from Apache Avro schemas
    id "application"
    id "com.commercehub.gradle.plugin.avro" version "0.8.0"
}

sourceCompatibility = '1.8'

mainClassName = 'plum.ExecutorApp'

repositories {
  mavenCentral()
}

version = '0.1.0'          Dependencies include
                           Kafka and Avro
dependencies {
  compile 'org.apache.kafka:kafka-clients:2.0.0'
  compile 'org.apache.avro:avro:1.8.2'
  compile 'net.sargue:mailgun:1.9.0'
  compile 'org.slf4j:slf4j-api:1.7.25'
}


jar {
  manifest {
    attributes 'Main-Class': mainClassName
  }

  from {
    configurations.compile.collect {
      it.isDirectory() ? it : zipTree(it)
    }
  } {
    exclude "META-INF/*.SF"
    exclude "META-INF/*.DSA"
    exclude "META-INF/*.RSA"
  }
}
```

Let's check that we can build this:

```
$ gradle compileJava
...
BUILD SUCCESSFUL
...
```

Now we are ready to work on the schema for our new command.

### 9.2.4 *Defining our alert schema*

To add our new schema for alerts to the project, create a file at this path:

src/main/resources/avro/alert.avsc

Populate this file with the Avro JSON schema in the following listing.

**Listing 9.2   alert.avsc**

```
{ "name": "Alert",
  "namespace": "plum.avro",
  "type": "record",
  "fields": [
    { "name": "command", "type": "string" },
    { "name": "notification",
      "type": {
        "name": "Notification",
        "namespace": "plum.avro",
        "type": "record",
        "fields": [
          { "name": "summary", "type": "string" },
          { "name": "detail", "type": "string" },
          { "name": "urgency",
            "type": {
              "type": "enum",
              "name": "Urgency",
              "namespace": "plum.avro",
              "symbols": ["HIGH", "MEDIUM", "LOW"]
            }
          }
        ]
      }
    },
    { "name": "recipient",
      "type": {
        "type": "record",
        "name": "Recipient",
        "namespace": "plum.avro",
        "fields": [
          { "name": "name", "type": "string" },
          { "name": "phone", "type": "string" },
          { "name": "email", "type": "string" }
        ]
      }
    },
    { "name": "createdAt", "type": "long" }
  ]
}
```

Quite a lot is going on in this Avro schema file. Let's break it down:

- Our top-level entity is a record called `Alert` that belongs in the `plum.avro` namespace (as do all of our entities).
- Our `Alert` consists of a `type` field, a `Notification` child record, a `Recipient` child record, and a `createdAt` timestamp for the alert.
- Our `Notification` record contains `summary`, `detail`, and `urgency` fields, where `urgency` is an enum with three possible values.
- Our `Recipient` record contains `name`, `phone`, and `email` fields.

A minor piece of housekeeping—we need to soft-link the resources/avro subfolder to another location so that the Avro plugin for Gradle can find it:

```
$ cd src/main && ln -s resources/avro .
```

With our Avro schema defined, let's use the Avro plugin in our build.gradle file to automatically generate Java bindings for our schema:

```
$ gradle generateAvroJava
:generateAvroProtocol UP-TO-DATE
:generateAvroJava

BUILD SUCCESSFUL

Total time: 8.234 secs
...
```

You will find the generated files inside your Gradle build folder:

```
$ ls build/generated-main-avro-java/plum/avro/
Alert.java  Notification.java  Recipient.java  Urgency.java
```

These files are too lengthy to reproduce here, but if you open them in your text editor, you will see that the files contain POJOs to represent the three records and one enumeration that make up our alert.

This is a wrap for designing our alert command for Plum and modeling that command in Apache Avro. Next, we can proceed to building our command executor.

## 9.3   Consuming our commands

Imagine that Plum now has a Kafka topic containing a constant stream of our alert commands, all stored in Avro format. With this stream in place, Plum now needs to implement a command executor, which will consume those alerts and execute them. First, we will get all the plumbing in place and check that we can successfully deserialize the incoming Avro event.

### 9.3.1   The right tool for the job

We could use lots of different stream-processing frameworks to execute our commands, but remember that command execution involves only two tasks:

1  Reading each command from the stream and executing it
2  Emitting an event to record that the command has been executed

Figure 9.5 shows the specifics of these two tasks for Plum.

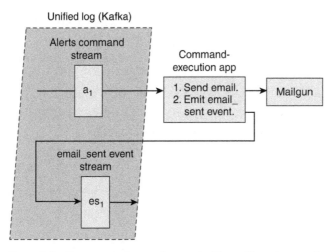

**Figure 9.5   The command-execution app for Plum will send an email
to the support engineer via Mailgun and then emit an `email_sent`
event to record that the alert has been executed.**

Both tasks are performed on one command at a time; there's no need to consider multiple commands at once. Our command execution is analogous to the single-event processing we explored in chapter 2. As with single-event processing, we will need only a simple framework to execute our commands, so the simple consumer and producer capabilities of the Kafka Java client library from chapter 2 will fit the bill nicely.

### 9.3.2   *Reading our commands*

As a first step, we need to read in our individual commands as records from our Kafka topic `commands`. Remember from chapter 2 that this is called a *consumer* in Kafka parlance. As in that chapter, we will write our own consumer in Java, using the Kafka Java client library.

Let's create a file for our consumer, called src/main/java/plum/Consumer.java, and add in the code in listing 9.3. This is a direct copy of the consumer code in chapter 2, except for the following changes:

- The new package name is `plum`.
- This code passes each consumed record to an `executor` rather than a `producer`.

**Listing 9.3   Consumer.java**

```
package plum;              ⟵┐  The package
                            │  is now plum.
import java.util.*;

import org.apache.kafka.clients.consumer.*;
```

```java
public class Consumer {

  private final KafkaConsumer<String, String> consumer;
  private final String topic;

  public Consumer(String servers, String groupId, String topic) {
    this.consumer = new KafkaConsumer<String, String>(
      createConfig(servers, groupId));
    this.topic = topic;
  }

  public void run(IExecutor executor) {
    this.consumer.subscribe(Arrays.asList(this.topic));
    while (true) {
      ConsumerRecords<String, String> records = consumer.poll(100);
      for (ConsumerRecord<String, String> record : records) {
        executor.execute(record.value());

      }
    }
  }

  private static Properties createConfig(
    String servers, String groupId) {

    Properties props = new Properties();
    props.put("bootstrap.servers", servers);
    props.put("group.id", groupId);
    props.put("enable.auto.commit", "true");
    props.put("auto.commit.interval.ms", "1000");
    props.put("auto.offset.reset", "earliest");
    props.put("session.timeout.ms", "30000");
    props.put("key.deserializer",
      "org.apache.kafka.common.serialization.StringDeserializer");
    props.put("value.deserializer",
      "org.apache.kafka.common.serialization.StringDeserializer");
    return props;
  }
}
```

**In place of the producer, we now have an executor.**

So far, so good—we have defined a consumer that will read all the records from a given Kafka topic and hand them over to the execute method of the supplied executor. Next, we will implement an initial version of the executor. This won't yet carry out the command, but it will show that we can successfully parse our Avro-structured events into Java objects.

### 9.3.3 Parsing our commands

See how our consumer is going to run the IExecutor.execute() method for each incoming command? To keep things flexible, the two executors we write in this chapter will both conform to the IExecutor interface, letting us easily swap out one for the other. Let's now define this interface in another file, called src/main/java/plum/ IExecutor.java. Add in the code in the following listing.

**Listing 9.4    IExecutor.java**

```java
package plum;

import java.util.Properties;

import org.apache.kafka.clients.producer.*;

public interface IExecutor {

  public void execute(String message);

  public static void write(KafkaProducer<String, String> producer,
    String topic, String message) {
    ProducerRecord<String, String> pr = new ProducerRecord(
      topic, message);
    producer.send(pr);
  }

  public static Properties createConfig(String servers) {
    Properties props = new Properties();
    props.put("bootstrap.servers", servers);
    props.put("acks", "all");
    props.put("retries", 0);
    props.put("batch.size", 1000);
    props.put("linger.ms", 1);
    props.put("key.serializer",
      "org.apache.kafka.common.serialization.StringSerializer");
    props.put("value.serializer",
      "org.apache.kafka.common.serialization.StringSerializer");
    return props;
  }
}
```

> Our abstract execute method, for concrete implementations of IExecutor to instantiate

Again, this code is extremely similar to the IProducer we wrote in chapter 2. As with that IProducer, this interface contains static helper methods to configure a Kafka record producer and write events to Kafka. We need these capabilities in our executor so that we can fulfill the second part of executing any given command: emitting an event back into Kafka and recording this command as having been successfully executed.

With the interface in place, let's write our first concrete implementation of IExecutor. At this stage, we won't execute the command, but we want to check that we can successfully parse the incoming command. Add the code in the following listing into a new file called src/main/java/plum/EchoExecutor.java.

**Listing 9.5    EchoExecutor.java**

```java
package plum;

import java.io.*;

import org.apache.kafka.clients.producer.*;
```

```
import org.apache.avro.*;
import org.apache.avro.io.*;
import org.apache.avro.generic.GenericData;
import org.apache.avro.specific.SpecificDatumReader;

import plum.avro.Alert;

public class EchoExecutor implements IExecutor {

  private final KafkaProducer<String, String> producer;
  private final String eventsTopic;

  private static Schema schema;
  static {
    try {
      schema = new Schema.Parser()
        .parse(EchoExecutor.class.getResourceAsStream("/avro/alert.avsc"));
    } catch (IOException ioe) {
      throw new ExceptionInInitializerError(ioe);
    }
  }

  public EchoExecutor(String servers, String eventsTopic) {

    this.producer = new KafkaProducer(IExecutor.createConfig(servers));
    this.eventsTopic = eventsTopic;
  }

  public void execute(String command) {

    InputStream is = new ByteArrayInputStream(command.getBytes());
    DataInputStream din = new DataInputStream(is);

    try {
      Decoder decoder = DecoderFactory.get().jsonDecoder(schema, din);
      DatumReader<Alert> reader = new SpecificDatumReader<Alert>(schema);
      Alert alert = reader.read(null, decoder);
      System.out.println("Alert " + alert.recipient.name + " about " +
        alert.notification.summary);
    } catch (IOException | AvroTypeException e) {
      System.out.println("Error executing command:" + e.getMessage());
    }
  }
}
```

**The Gradle Avro plugin automatically generates this Alert POJO from the bundled schema.**

**Statically initialize this Schema object from the bundled schema.**

**Deserialize our alert command into an Alert POJO.**

**Print out salient information about the alert.**

The EchoExecutor is simple. Every time the execute method is invoked with a serialized command, it does the following:

1 Attempts to deserialize the incoming command into an Alert POJO, where that Alert POJO has been statically generated (by the Gradle Avro plugin) from the alert.avsc schema

2 If successful, prints the salient information about the alert out to stdout

### 9.3.4    *Stitching it all together*

We can now stitch these three files together via a new ExecutorApp class containing our main method. Create a new file called src/main/java/plum/ExecutorApp.java and populate it with the contents of the following listing.

**Listing 9.6    ExecutorApp.java**

```java
package plum;

public class ExecutorApp {

  public static void main(String[] args){
    String servers      = args[0];
    String groupId      = args[1];
    String commandsTopic = args[2];
    String eventsTopic   = args[3];

    Consumer consumer = new Consumer(servers, groupId, commandsTopic);
    EchoExecutor executor = new EchoExecutor(servers, eventsTopic);
    consumer.run(executor);
  }
}
```

We will pass four arguments into our StreamApp on the command line:

- servers specifies the host and port for talking to Kafka.
- groupId identifies our code as belonging to a specific Kafka consumer group.
- commandsTopic is the Kafka topic of commands to read from.
- eventsTopic is the Kafka topic we will write events to.

Let's build our stream processing app now. From the project root, the plum folder, run this:

```
$ gradle jar
...
BUILD SUCCESSFUL

Total time: 25.532 secs
```

We are now ready to test our stream processing app.

### 9.3.5    *Testing*

To test our new application, we are going to need five terminal windows. Figure 9.6 sets out what we'll be running in each of these terminals.

Each of our first three terminal windows will run a shell script from inside our Kafka installation directory:

```
$ cd ~/kafka_2.12-2.0.0
```

**Figure 9.6   The five terminals we need to run in order to test our command-executor app include ZooKeeper, Kafka, one command producer, the command executor app, and a consumer for the events emitted by the command executor.**

In our first terminal, we start up ZooKeeper:

```
$ bin/zookeeper-server-start.sh config/zookeeper.properties
```

In our second terminal, we start up Kafka:

```
$ bin/kafka-server-start.sh config/server.properties
```

In our third terminal, let's start a script that lets us send commands into our `alerts` Kafka topic:

```
$ bin/kafka-console-producer.sh --topic alerts \
  --broker-list localhost:9092
```

Let's now give this producer an alert command, by pasting this into the same terminal, making sure to add a newline after the command to send it into the Kafka topic:

```
{ "command" : "alert", "notification": { "summary": "Overheating machine",
  "detail": "Machine cz1-123 may be overheating!", "urgency": "MEDIUM" },
  "recipient": { "name": "Suzie Smith", "phone": "(541) 754-3010", "email":
  "s.smith@plum.com" }, "createdAt": 1539576669992 }
```

Phew! We are finally ready to start up our new command-executing application. In a fourth terminal, head back to your project root, the plum folder, and run this:

```
$ cd ~/plum
$ java -jar ./build/libs/plum-0.1.0.jar localhost:9092 ulp-ch09 \
  alerts events
```

This has kicked off our app, which will now read all commands from `alerts` and execute them. Wait a second and you should see the following output in the same terminal:

```
Alert Suzie Smith about Overheating machine
```

Good news: our simple `EchoExecutor` is working a treat. Now we can move onto the more complex version to execute the alert and log the sent email. Shut down the stream processing app with Ctrl-Z and then type `kill %%`, but make sure to leave that terminal and the other terminal windows open for the next section.

## 9.4    *Executing our commands*

Now that we can successfully parse our alert commands from Avro-based JSONs into POJOs, we can now move on to *executing* those commands. In this section, we will wire in a third-party email service to send our alerts, and then finish up by emitting an event to record that the email has been sent. Let's get started.

### 9.4.1   *Signing up for Mailgun*

Our overlords at Plum want the monitoring alerts about the NCX-10 machines to be sent to the maintenance engineers via email. So, we need to integrate some kind of email-sending mechanism into our executor. We have many to choose from out there, but we will go with a hosted transactional email service called Mailgun (www .mailgun.com).

Mailgun lets us sign up for a new account and start sending emails without providing any billing information, which is perfect for this kind of experimentation. If you prefer, you could equally use an alternative provider like Amazon Simple Email Service (SES), SendGrid, or even a local mail server option such as Postfix. If you go with an alternative email service provider, you will need to adjust the following instructions accordingly.

Head to the signup page for Mailgun: https://signup.mailgun.com/new/signup.

Fill in your details; you don't need to provide any billing information. Click the Create Account button, and on the next screen you will see some Java code for sending an email, as shown in figure 9.7. Before you can send an email, you have to do two more things:

1   Activate your Mailgun account. Mailgun will have sent a confirmation email to your signup email; click on this and follow the instructions to authenticate your account.
2   Add an authorized recipient. You can follow the link at the bottom of figure 9.7 to add an email recipient for testing. If you don't have a second email address, you can add *+ulp* or something similar to your existing address (for example, alex+ulp@foo.com). Again, you need to click on the confirmation email to authorize this.

### 9.4.2   *Completing our executor*

With these steps completed, we are now ready to update our executor to send emails. First, we need a thin wrapper around the Mailgun email-sending functionality: create a file called src/main/java/plum/Emailer.java and add in the contents of the following listing.

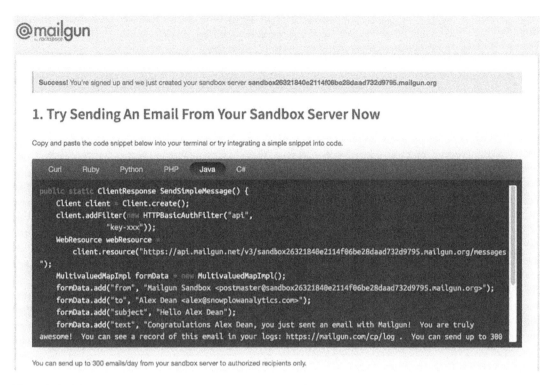

**Figure 9.7  Click the Java tab on this get-started screen from Mailgun and you'll see some basic code for sending an email via Mailgun from Java.**

Listing 9.7  Emailer.java

```java
package plum;

import net.sargue.mailgun.*;

import plum.avro.Alert;

public final class Emailer {

  static final String MAILGUN_KEY =
    "XXX";
  static final String MAILGUN_SANDBOX =
    "sandboxYYY.mailgun.org";

  private static final Configuration configuration = new Configuration()
    .domain(MAILGUN_SANDBOX)
    .apiKey(MAILGUN_KEY)
    .from("Test account", "postmaster@" + MAILGUN_SANDBOX);

  public static void send(Alert alert) {
    Mail.using(configuration)
```

Replace the XXX with your Mailgun API key.

Replace the YYY with your Mailgun sandbox server.

```
          .to(alert.recipient.email.toString())
          .subject(alert.notification.summary.toString())
          .text(alert.notification.detail.toString())
          .build()
          .send();
   }
}
```

Emailer.java contains a simple wrapper around the Mailgun emailing library that we are using. Make sure to update the constants with the API key and sandbox server details found in your Mailgun account.

With this in place, we can now write our full executor. Like the EchoExecutor, this will implement the IExecutor interface, but the FullExecutor will also send the email via Mailgun and log the event to our events topic in Kafka. Add the contents of the following listing into the file src/main/java/plum/FullExecutor.java.

**Listing 9.8  FullExecutor.java**

```
package plum;

import java.io.*;

import org.apache.kafka.clients.producer.*;

import org.apache.avro.*;
import org.apache.avro.io.*;
import org.apache.avro.generic.GenericData;
import org.apache.avro.specific.SpecificDatumReader;

import plum.avro.Alert;

public class FullExecutor implements IExecutor {

  private final KafkaProducer<String, String> producer;
  private final String eventsTopic;

  private static Schema schema;
  static {
    try {
      schema = new Schema.Parser()
        .parse(EchoExecutor.class.getResourceAsStream("/avro/alert.avsc"));
    } catch (IOException ioe) {
      throw new ExceptionInInitializerError(ioe);
    }
  }

  public FullExecutor(String servers, String eventsTopic) {

    this.producer = new KafkaProducer(IExecutor.createConfig(servers));
    this.eventsTopic = eventsTopic;
  }

  public void execute(String command) {
```

```
                InputStream is = new ByteArrayInputStream(command.getBytes());
                DataInputStream din = new DataInputStream(is);

                try {
                  Decoder decoder = DecoderFactory.get().jsonDecoder(schema, din);
                  DatumReader<Alert> reader = new SpecificDatumReader<Alert>(schema);
                  Alert alert = reader.read(null, decoder);
                  Emailer.send(alert);
                  IExecutor.write(this.producer, this.eventsTopic,
                    "{ \"event\": \"email_sent\" }");
                } catch (IOException | AvroTypeException e) {
                  System.out.println("Error executing command:" + e.getMessage());
                }
              }
            }
          }
```

**Send our email via Mailgun.** → `Emailer.send(alert);`

**Emit an event to record the event being sent.** → `IExecutor.write(this.producer, this.eventsTopic,`

FullExecutor is similar to our earlier EchoExecutor, but with two additions:

- We are now sending the email to our support engineer via the new Emailer.
- Following the successful email send, we are logging a basic event to record that success to a Kafka topic containing events.

Now we need to rewrite the entry point to our app to use our new executor. Edit the file src/main/java/plum/ExecutorApp.java and repopulate it with the contents of the following listing.

**Listing 9.9 ExecutorApp.java**

```
package plum;

import java.util.Properties;

public class ExecutorApp {

  public static void main(String[] args){
    String servers      = args[0];
    String groupId      = args[1];
    String commandsTopic = args[2];
    String eventsTopic   = args[3];

    Consumer consumer = new Consumer(servers, groupId, commandsTopic);
    FullExecutor executor = new FullExecutor(servers, eventsTopic);      ←┐
    consumer.run(executor);
  }
}
```

**Replaced the EchoExecutor with a FullExecutor**

Let's rebuild our app now. From the project root, the plum folder, run this:

```
$ gradle jar
...
BUILD SUCCESSFUL

Total time: 25.532 secs
```

We are now ready to rerun our command executor.

### 9.4.3    *Final testing*

Head back to the terminal running the previous build of the executor. If it is still running, kill it with Ctrl-C. Now restart it:

```
$ java -jar ./build/libs/plum-0.1.0.jar localhost:9092 ulp-ch09 \
  alerts events
```

Leave ZooKeeper and Kafka running in their respective terminals; we now need to start the fifth and final terminal, to tail our Kafka topic containing events:

```
$ bin/kafka-console-consumer.sh --topic events --from-beginning \
  --bootstrap-server localhost:9092
```

Now let's head back to the terminal that is running a producer connected to our alerts topic. We'll take the alert command we ran before, update it so that the recipient's email is the same one that we authorized with Mailgun, and then paste it into the producer:

```
$ bin/kafka-console-producer.sh --topic alerts \
  --broker-list localhost:9092
{ "command" : "alert", "notification": { "summary": "Overheating machine",
 "detail": "Machine cz1-123 may be overheating!", "urgency": "MEDIUM" },
 "recipient": { "name": "Suzie Smith", "phone": "(541) 754-3010", "email":
 "alex+test@snowplowanalytics.com" }, "createdAt": 1543392786232 }
```

For this to work, you *must* update the email in the alert to your own email, the one that you authorized with Mailgun. Check back in your new terminal that is tailing our Kafka events topic and you should see this:

```
$ bin/kafka-console-consumer.sh --topic events --from-beginning \
  --zookeeper localhost:2181
{ "event": "email_sent" }
```

That's encouraging; it suggests that our command executor has sent our email. Check in your email client and you should see an incoming email, something like figure 9.8.

**Figure 9.8    Our command-executor app is now able to email alert commands to us via Mailgun.**

Great! We have now implemented a command executor that can take incoming commands—in our case, alerts for a long-suffering Plum maintenance engineer—and convert them into emails to that engineer. Our command executor even emits an email_sent event to track that the action has been performed.

## 9.5     Scaling up commands

This chapter has walked you through a simple example of executing an alert command as it passes through Plum's unified log. So far, so good—but how do we scale this up to a real company with hundreds or thousands of possible commands? This section has some ideas.

### 9.5.1     One stream of commands, or many?

In this chapter, we called our stream of commands `alerts` on the basis that it would contain only alert commands, and our command-executor app processed all incoming commands as alerts. If we want to extend our implementation to support hundreds or thousands of commands, we have several options:

- Have one stream (topic, in Kafka parlance) per command—in other words, hundreds or thousands of streams.
- Make the commands self-describing but write them to say, three or five streams; each stream represents a different command *priority*.
- Make the commands *self-describing* and have a single stream. Include a header in each record that tells the command executor what type of command this is.

Figure 9.9 illustrates these three options.

When choosing one of these options, you will want to consider the operational overhead of having multiple streams versus the development overhead of making your commands self-describing. It is also important to understand how these different setups fare in the face of *command-execution failures*.

### 9.5.2     Handling command-execution failures

Imagine that Mailgun has an outage (whether planned or unplanned) for several hours. What will happen to our command executor? Remember that our command executor is performing single-event processing; it has no way of knowing that a systemic problem has arisen with Mailgun. If we are lucky, our Mailgun email-sending code will simply time out after perhaps 30 seconds, for each alert command that is processed. In this case, we would likely emit an Email Failed to Send event or similar back into Kafka.

Now imagine that our command executor is reading a stream containing many types of commands. The failure of each alert command costs us 30 seconds of processing time, so if our decision-making apps are generating new commands at a faster rate than this, our command executor will fall further and further behind. Other high-priority commands that Plum wants to execute will be extremely delayed.

Note that the sharded nature of a Kafka or Kinesis stream doesn't help you: assuming that the failing command type is distributed across all of the shards, the worker assigned to each shard is equally likely to suffer slowdowns as the other workers. This is illustrated in figure 9.10.

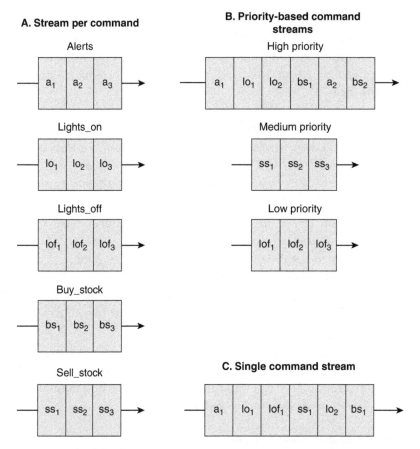

**Figure 9.9   We could define one stream for each command type (option A), associate commands to priority-based streams (option B), or use one stream for all commands (option C).**

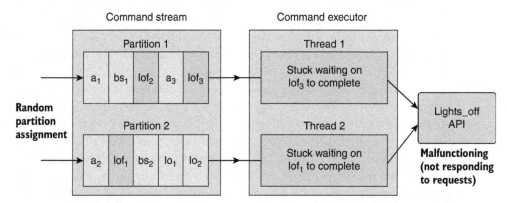

**Figure 9.10   With a command stream containing two partitions, a single malfunctioning command can cause both threads within the command executor (one per partition) to get "stuck." Here, the `lights_off` command is failing to execute because of the unavailability of the Lights_off external API.**

Separating the commands into streams with different priorities does not necessarily help either: it is still possible for a high-priority command type to slow other high-priority commands, as also seen in figure 9.10.

Two potential solutions to consider are as follows:

- Having a separate command executor for all high- or medium-priority command types, so that each command executor operates (and fails) independently of the others.
- Copying all commands into a separate publish-subscribe message queue, such as NSQ (see chapter 1) or SQS, with a scalable pool of workers to execute the commands. This queue would not have the Kafka or Kinesis notion of ordered processing, so there would less scope for functioning command types to be "blocked" by failing commands.

### 9.5.3  *Command hierarchies*

In this chapter, we have argued in favor of creating commands that are *shrink-wrapped* and *fully baked*. A command executor should have discretion as to how each command should be best executed, but the executor should not have to look up additional data or contain command-specific business logic. Following this design, we created an alert command for Plum, which was resolved by our executor into an email sent via Mailgun.

The alert command is a useful one, but under the hood our command executor was resolving it into a *send email* command, and then ultimately into a *send email via Mailgun* command. For your unified log, you may want to model these more granular commands as well as the high-level ones, so that a decision-making app can be as precise or flexible as it likes as to the commands it emits. Figure 9.11 illustrates this idea of a form of *command hierarchy*.

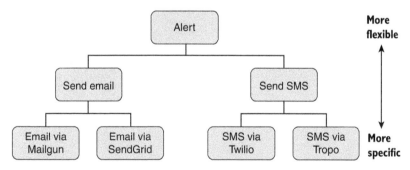

**Figure 9.11  In a hierarchy of commands, each generation is more specific than the one above it. Under alert, we have a choice between alerting via email and alerting via SMS; within email and SMS, we have even more-specific commands to determine which provider is used for the sending.**

## *Summary*

- A command is an order or instruction for a specific action to be performed in the future. Each command, when executed, produces an event.

- Whereas most software uses implicit decision-making, apps in a unified log architecture can perform explicit decision-making that emits commands to one or more streams within the unified log.

- A stream of commands in a unified log can be executed using dedicated applications performing single-event processing.

- Commands should be carefully modeled to ensure that they are shrink-wrapped and fully baked. This ensures that decision-making and command execution can stay loosely coupled, with a clear separation of concerns.

- Commands can be modeled in Apache Avro or another schema tool (see chapter 6).

- Command executor apps can be implemented using simple Kafka consumer and producer patterns in Java.

- On successful execution of a command, the executor should emit an event recording the exact action performed.

- When scaling up an implementation of commands in your unified log, consider how many command streams to operate, how to handle execution failures, and whether to implement command hierarchies.

# Part 3

# Event analytics

In this last part of the book, we'll introduce two well-known techniques for performing analytics on event streams—namely, analytics-on-write and analytics-on-read. Using illustrative examples, we'll explain which one to use depending on the circumstances.

# Analytics-on-read

**This chapter covers**

- Analytics-on-read versus analytics-on-write
- Amazon Redshift, a horizontally scalable columnar database
- Techniques for storing and widening events in Redshift
- Some example analytics-on-read queries

Up to this point, this book has largely focused on the operational mechanics of a unified log. When we have performed any analysis on our event streams, this has been primarily to put technology such as Apache Spark or Samza through its paces, or to highlight the various use cases for a unified log.

Part 3 of this book sees a change of focus: we will take an analysis-first look at the unified log, leading with the two main methodologies for *unified log analytics* and then applying various database and stream processing technologies to analyze our event streams.

What do we mean by *unified log analytics?* Simply put, unified log analytics is the examination of one or more of our unified log's event streams to drive business value. It covers everything from detection of customer fraud, through KPI dashboards for busy executives, to predicting breakdowns of fleet vehicles or plant machinery. Often the consumer of this analysis will be a human or humans, but not necessarily: unified log analytics can just as easily drive an automated machine-to-machine response.

With unified log analytics having such a broad scope, we need a way of breaking down the topic further. A helpful distinction for analytics on event streams is between *analytics-on-write* versus *analytics-on-read*. The first section of this chapter will explain the distinction, and then we will dive into a complete case study of analytics-on-read. A new part of the book deserves a new figurative unified log owner, so for our case study we will introduce OOPS, a major package-delivery company.

Our case study will be built on top of the Amazon Redshift database. *Redshift* is a fully hosted (on Amazon Web Services) analytical database that uses columnar storage and speaks a variant of PostgreSQL. We will be using Redshift to try out a variety of analytics-on-read required by OOPS.

Let's get started!

## 10.1  *Analytics-on-read, analytics-on-write*

If we want to explore what is occurring in our event streams, where should we start? The Big Data ecosystem is awash with competing databases, batch- and stream-processing frameworks, visualization engines, and query languages. Which ones should we pick for our analyses?

The trick is to understand that all these myriad technologies simply help us to implement either *analytics-on-write* or *analytics-on-read* for our unified log. If we can understand these two approaches, the way we deliver our analytics by using these technologies should become much clearer.

### 10.1.1  *Analytics-on-read*

Let's imagine that we are in the BI team at OOPS, an international package-delivery company. OOPS has implemented a unified log using Amazon Kinesis that is receiving events emitted by OOPS delivery trucks and the handheld scanners of OOPS delivery drivers.

We know that as the BI team, we will be responsible for analyzing the OOPS event streams in all sorts of ways. We have some ideas of what reports to build, but these are just hunches until we are much more familiar with the events being generated by OOPS trucks and drivers out in the field. How can we get that familiarity with our event streams? This is where analytics-on-read comes in.

*Analytics-on-read* is really shorthand for a two-step process:

1  *Write* all of our events to some kind of event store.
2  *Read* the events from our event store to perform an analysis.

In other words: store first; ask questions later. Does this sound familiar? We've been here before, in chapter 7, where we archived all of our events to Amazon S3 and then wrote a simple Spark job to generate a simple analysis from those events. This was classic *analytics-on-read*: first we wrote our events to a storage target (S3), and only later did we perform the required analysis, when our Spark job read all of our events back from our S3 archive.

An analytics-on-read implementation has three key parts:

- A *storage target* to which the events will be written. We use the term *storage target* because it is more general than the term *database*.
- A *schema, encoding*, or *format* in which the events should be written to the storage target.
- A *query engine* or *data processing framework* to allow us to analyze the events as read from our storage target.

Figure 10.1 illustrates these three components.

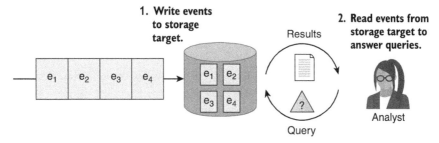

**Figure 10.1   For analytics-on-read, we first write events to our storage target in a predetermined format. Then when we have a query about our event stream, we run that query against our storage target and retrieve the results.**

Storing data in a database so that it can be queried later is a familiar idea to almost all readers. And yet, in the context of the unified log, this is only half of the analytics story; the other half of the story is called *analytics-on-write*.

### 10.1.2   Analytics-on-write

Let's skip forward in time and imagine that we, the BI team at OOPS, have implemented some form of analytics-on-read on the events generated by our delivery trucks and drivers. The exact implementation of this analytics doesn't particularly matter. For familiarity's sake, let's say that we again stored our events in Amazon S3 as JSON, and we then wrote a variety of Apache Spark jobs to derive insights from the event archive.

Regardless of the technology, the important thing is that as a team, we are now much more comfortable with the contents of our company's unified log. And this familiarity has spread to other teams within OOPS, who are now making their own demands on our team:

- The *executive team* wants to see *dashboards of key performance indicators (KPIs)*, based on the event stream and accurate to the last five minutes.

- The *marketing team* wants to add a *parcel tracker on the website,* using the event stream to show customers where their parcel is right now and when it's likely to arrive.
- The *fleet maintenance team* wants to use simple algorithms (such as date of last oil change) to identify trucks that may be about to break down in mid-delivery.

Your first thought might be that our standard analytics-on-read could meet these use cases; it may well be that your colleagues have prototyped each of these requirements with a custom-written Spark job running on our archive of events in Amazon S3. But putting these three reports into production will take an analytical system with different priorities:

- *Very low latency*—The various dashboards and reports must be fed from the incoming event streams in as close to real time as possible. The reports must not lag more than five minutes behind the present moment.
- *Supports thousands of simultaneous users*—For example, the parcel tracker on the website will be used by large numbers of OOPS customers at the same time.
- *Highly available*—Employees and customers alike will be depending on these dashboards and reports, so they need to have excellent uptime in the face of server upgrades, corrupted events, and so on.

These requirements point to a much more operational analytical capability—one that is best served by *analytics-on-write.* Analytics-on-write is a four-step process:

1  *Read* our events from our event stream.
2  *Analyze* our events by using a stream processing framework.
3  *Write* the summarized output of our analysis to a storage target.
4  *Serve* the summarized output into real-time dashboards or reports.

We call this *analytics-on-write* because we are performing the analysis portion of our work prior to writing to our storage target; you can think of this as *early,* or *eager,* analysis, whereas analytics-on-read is *late,* or *lazy,* analysis. Again, this approach should seem familiar; when we were using Apache Samza in part 1, we were practicing a form of analytics-on-write!

Figure 10.2 shows a simple example of analytics-on-write using a key-value store.

### 10.1.3  *Choosing an approach*

As the OOPS BI team, how should we choose between analytics-on-read and analytics-on-write? Strictly speaking, we don't have to choose, as illustrated in figure 10.3. We can attach multiple analytics applications, both read and write, to process the event streams within OOPS's unified log.

Most organizations, however, will start with *analytics-on-read.* Analytics-on-read lets you explore your event data in a flexible way: because you have all of your events stored and a query language to interrogate them, you can answer pretty much any question

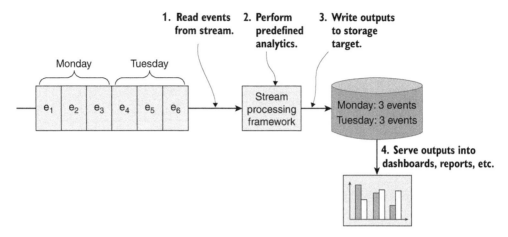

**Figure 10.2** With analytics-on-write, the analytics are performed in stream, typically in close to real time, and the outputs of the analytics are written to the storage target. Those outputs can then be served into dashboards and reports.

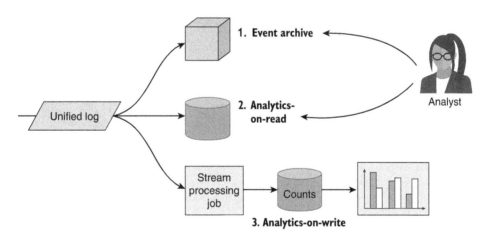

**Figure 10.3** Our unified log feeds three discrete systems: our event archive, an analytics-on-read system, and an analytics-on-write system. Event archives were discussed in chapter 7.

asked of you. Specific analytics-on-write requirements will likely come later, as this initial analytical understanding percolates through your business.

In chapter 11, we will explore analytics-on-write in more detail, which should give you a better sense of when to use it; in the meantime, table 10.1 sets out some of the key differences between the two approaches.

**Table 10.1    Comparing the main attributes of analytics-on-read to analytics-on-write**

| Analytics-on-read | Analytics-on-write |
| --- | --- |
| Predetermined storage format | Predetermined storage format |
| Flexible queries | Predetermined queries |
| High latency | Low latency |
| Support 10–100 users | Support 10,000s of users |
| Simple (for example, HDFS) or sophisticated (for example, HP Vertica) storage target | Simple storage target (for example, key-value store) |
| Sophisticated query engine or batch processing framework | Simple (for example, AWS Lambda) or sophisticated (for example, Apache Samza) stream processing framework |

## 10.2    *The OOPS event stream*

It's our first day on the OOPS BI team, and we have been asked to familiarize ourselves with the various event types being generated by OOPS delivery trucks and drivers. Let's get started.

### 10.2.1    *Delivery truck events and entities*

We quickly learn that three types of events are related to the delivery trucks themselves:

- Delivery truck departs from location at time
- Delivery truck arrives at location at time
- Mechanic changes oil in delivery truck at time

We ask our new colleagues about the three entities involved in our events: *delivery trucks*, *locations*, and *employees*. They talk us through the various properties that are stored against each of these entities when the event is emitted by the truck. Figure 10.4 illustrates these properties.

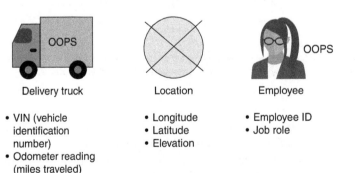

Delivery truck

- VIN (vehicle identification number)
- Odometer reading (miles traveled)

Location

- Longitude
- Latitude
- Elevation

Employee

- Employee ID
- Job role

**Figure 10.4    The three entities represented in our delivery-truck events are the delivery truck itself, a location, and an employee. All three of these entities have minimal properties—just enough to uniquely identify the entity.**

Note how few properties there are in these three event entities. What if we want to know more about a specific delivery truck or the employee changing its oil? Our colleagues in BI assure us that this is possible: they have access to the OOPS vehicle database and HR systems, containing data stored against the vehicle identification number (VIN) and the employee ID. Hopefully, we can use this additional data later.

### 10.2.2 Delivery driver events and entities

Now we turn to the delivery drivers themselves. According to our BI team, just two events matter here:

- Driver delivers package to customer at location at time
- Driver cannot find customer for package at location at time

In addition to employees and locations, which we are familiar with, these events involve two new entity types: *packages* and *customers*. Again, our BI colleagues talk us through the properties attached to these two entities in the OOPS event stream. As you can see in figure 10.5, these entities have few properties—just enough data to uniquely identify the package or customer.

Package
- Package ID

Customer
- Customer ID
- Is VIP?

**Figure 10.5  The events generated by our delivery drivers involve two additional entities: packages and customers. Again, both entities have minimal properties in the OOPS event model.**

### 10.2.3 The OOPS event model

The event model used by OOPS is slightly different from the one we used in parts 1 and 2 of the book. Each OOPS event is expressed in JSON and has the following properties:

- A short tag describing the event (for example, TRUCK_DEPARTS or DRIVER_MISSES _CUSTOMER)
- A timestamp at which the event took place
- Specific slots related to each of the entities that are involved in this event

Here is an example of a *Delivery truck departs from location at time* event:

```
{ "event": "TRUCK_DEPARTS", "timestamp": "2018-11-29T14:48:35Z",
 "vehicle": { "vin": "1HGCM82633A004352", "mileage": 67065 }, "location":
 { "longitude": 39.9217860, "latitude": -83.3899969, "elevation": 987 } }
```

The BI team at OOPS has made sure to document the structure of all five of their event types by using JSON Schema; remember that we introduced JSON Schema in chapter 6. In the following listing, you can see the JSON schema file for the *Delivery truck departs* event.

---

**Listing 10.1   truck_departs.json**

```json
{
  "type": "object",
  "properties": {
    "event": {
      "enum": [ "TRUCK_DEPARTS" ]
    },
    "timestamp": {
      "type": "string",
      "format": "date-time"
    },
    "vehicle": {
      "type": "object",
      "properties": {
        "vin": {
          "type": "string",
          "minLength": 17,
          "maxLength": 17
        },
        "mileage": {
          "type": "integer",
          "minimum": 0,
          "maximum": 2147483647
        }
      },
      "required": [ "vin", "mileage" ],
      "additionalProperties": false
    },
    "location": {
      "type": "object",
      "properties": {
        "latitude": {
          "type": "number"
        },
        "longitude": {
          "type": "number"
        },
        "elevation": {
          "type": "integer",
          "minimum": -32768,
          "maximum": 32767
        }
      },
      "required": [ "longitude", "latitude", "elevation" ],
      "additionalProperties": false
    }
  },
  "required": [ "event", "timestamp", "vehicle", "location" ],
  "additionalProperties": false
}
```

### 10.2.4 *The OOPS events archive*

By the time we join the BI team, the unified log has been running at OOPS for several months. The BI team has implemented a process to archive the events being generated by the OOPS delivery trucks and drivers. This archival approach is similar to that taken in chapter 7:

1 Our delivery trucks and our drivers' handheld computers emit events that are received by an event collector (most likely a simple web server).
2 The event collector writes these events to an event stream in Amazon Kinesis.
3 A stream processing job reads the stream of raw events and validates the events against their JSON schemas.
4 Events that fail validation are written to a second Kinesis stream, called the *bad stream*, for further investigation.
5 Events that pass validation are written to an Amazon S3 bucket, and partitioned into folders based on the event type.

Phew—there is a lot to take in here, and you may be thinking that these steps are not relevant to us, because they take place upstream of our analytics. But a good analyst or data scientist will always take the time to understand the source-to-sink lineage of their event stream. We should be no different! Figure 10.6 illustrates this five-step event-archiving process.

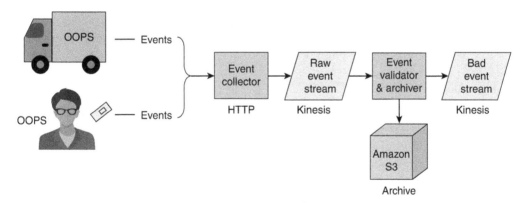

**Figure 10.6 Events flow from our delivery trucks and drivers into an event collector. The event collector writes the events into a raw event stream in Kinesis. A stream processing job reads this stream, validates the events, and archives the valid events to Amazon S3; invalid events are written to a second stream.**

A few things to note before we continue:

- The validation step is important, because it ensures that the archive of events in Amazon S3 consists only of well-formed JSON files that conform to JSON Schema.
- In this chapter, we won't concern ourselves further with the bad stream, but see chapter 8 for a more thorough exploration of happy versus failure paths.

- The events are stored in Amazon S3 in uncompressed plain-text files, with each event's JSON separated by a newline. This is called *newline-delimited JSON* (http://ndjson.org).

To check the format of the events ourselves, we can download them from Amazon S3:

```
$ aws s3 cp s3://ulp-assets-2019/ch10/data/ . --recursive --profile=ulp
download: s3://ulp-assets-2019/ch10/data/events.ndjson to ./events.ndjson
$ head -3 events.ndjson
{"event":"TRUCK_DEPARTS", "location":{"elevation":7, "latitude":51.522834,
 "longitude": -0.081813}, "timestamp":"2018-11-01T01:21:00Z", "vehicle":
 {"mileage":32342, "vin":"1HGCM82633A004352"}}
{"event":"TRUCK_ARRIVES", "location":{"elevation":4, "latitude":51.486504,
 "longitude": -0.0639602}, "timestamp":"2018-11-01T05:35:00Z", "vehicle":
 {"mileage":32372, "vin":"1HGCM82633A004352"}}
{"employee":{"id":"f6381390-32be-44d5-9f9b-e05ba810c1b7", "jobRole":
 "JNR_MECHANIC"}, "event":"MECHANIC_CHANGES_OIL", "timestamp":
 "2018-11-01T08:34:00Z", "vehicle": {"mileage":32372, "vin":
 "1HGCM82633A004352"}}
```

We can see some events that seem to relate to a trip to the garage to change a delivery truck's oil. And it looks like the events conform to the JSON schema files written by our colleagues, so we're ready to move onto Redshift.

## 10.3 Getting started with Amazon Redshift

Our colleagues in BI have selected Amazon Redshift as the storage target for our analytics-on-read endeavors. In this section, we will help you become familiar with Redshift before designing an event model to store our various event types in the database.

### 10.3.1 Introducing Redshift

Amazon Redshift is a column-oriented database from Amazon Web Services that has grown increasingly popular for event analytics. Redshift is a fully hosted database, available exclusively on AWS and built using columnar database technology from ParAccel (now part of Actian). ParAccel's technology is based on PostgreSQL, and you can largely use PostgreSQL-compatible tools and drivers to connect to Redshift.

Redshift has evolved significantly since its launch in early 2013, and now sports distinct features of its own; even if you are familiar with PostgreSQL or ParAccel, we recommend checking out the official Redshift documentation from AWS at https://docs.aws.amazon.com/redshift/.

Redshift is a massively parallel processing (MPP) database technology that allows you to scale horizontally by adding additional *nodes* to your *cluster* as your event volumes grow. Each Redshift cluster has a leader node and at least one compute node. The leader node is responsible for receiving SQL queries from a client, creating a query execution plan, and then farming out that query plan to the compute nodes. The compute nodes execute the query execution plan and then return their portion of the results back to the leader node. The leader node then consolidates the results from all of the compute nodes and returns the final results to the client. Figure 10.7 illustrates this data flow.

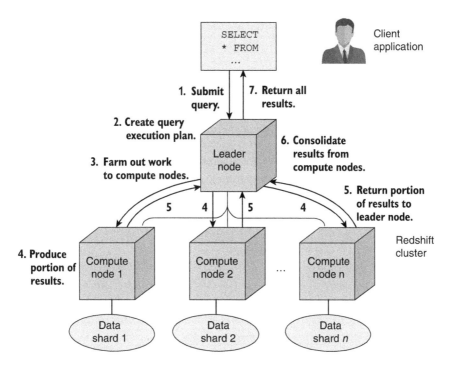

**Figure 10.7  A Redshift cluster consists of a leader node and at least one compute node. Queries flow from a client application through the leader node to the compute nodes; the leader node is then responsible for consolidating the results and returning them to the client.**

Redshift has attributes that make it a great fit for many analytics-on-read workloads. Its killer feature is the ability to run PostgreSQL-flavored SQL, including full inter-table JOINs and windowing functions, over many billions of events. But it's worth understanding the design decisions that have gone into Redshift, and how those decisions enable some things while making other things more difficult. Table 10.2 sets out these related strengths and weaknesses of Redshift.

**Table 10.2  Strengths and weaknesses of Amazon Redshift**

| Redshift's strengths | | Redshift's weaknesses |
|---|---|---|
| Scales horizontally to many billions of events. | *but* | High latency for even simple query responses. |
| Columnar storage allows for fast aggregations for data modeling and reporting. | *but* | Retrieving rows (for example, SELECT * FROM …) is slow. |
| Append-only data loading from Amazon S3 or DynamoDB is fast. | *but* | Updating existing data is painful. |

**Table 10.2   Strengths and weaknesses of Amazon Redshift (continued)**

| Redshift's strengths | | Redshift's weaknesses |
|---|---|---|
| Can prioritize important queries and users via workload management (WLM). | *but* | Designed for tens of users, not hundreds or thousands. |

In any case, our BI colleagues at OOPS have asked us to set up a new Redshift cluster for some analytics-on-write experiments, so let's get started!

### 10.3.2  Setting up Redshift

Our colleagues haven't given us access to the OOPS AWS account yet, so we will have to create the Redshift cluster in our own AWS account. Let's do some research and select the smallest and cheapest Redshift cluster available; we must remember to shut it down as soon as we are finished with it as well.

At the time of writing, four Amazon Elastic Compute Cloud (EC2) instance types are available for a Redshift cluster, as laid out in table 10.3.[1]

A cluster must consist exclusively of one instance type; you cannot mix and match. Some instance types allow for single-instance clusters, whereas other instance types require at least two instances. Confusingly, AWS refers to this as *single-node* versus *multi-node*—confusing because a single-node cluster still has a leader node and a compute node, but they simply reside on the same EC2 instance. On a multi-node cluster, the leader node is separate from the compute nodes.

**Table 10.3   Four EC2 instance types available for a Redshift cluster**

| Instance type | Storage | CPU | Memory | Single-instance okay? |
|---|---|---|---|---|
| dc2.large | 160 GB SSD | 7 EC2 CUs | 15 GiB | Yes |
| dc2.8xlarge | 2.56 TB SSD | 99 EC2 CUs | 244 GiB | No |
| ds2.xlarge | 2 TB HDD | 14 EC2 CUs | 31 GiB | Yes |
| ds2.8xlarge | 16 TB HDD | 116 EC2 CUs | 244 GiB | No |

Currently, you can try Amazon Redshift for two months for free (provided you've never created an Amazon Redshift cluster before) and you get a cluster consisting of a single dc2.large instance, so that's what we will set up using the AWS CLI. Assuming you are working in the book's Vagrant virtual machine and still have your ulp profile set up, you can simply type this:

---

[1]   In table 10.3, the "Instance type" column uses "dc" and "ds" prefixes to refer to dense compute and dense storage, respectively. In the "CPU" column, "CUs" refers to compute units.

```
$ aws redshift create-cluster --cluster-identifier ulp-ch10 \
--node-type  dc2.large --db-name ulp --master-username ulp \
--master-user-password Unif1edLP --cluster-type single-node \
--region us-east-1 --profile ulp
```

As you can see, creating a new cluster is simple. In addition to specifying the cluster type, we create a master username and password, and create an initial database called ulp. Press Enter, and the AWS CLI should return JSON of the new cluster's details:

```
{
    "Cluster": {
        "ClusterVersion": "1.0",
        "NumberOfNodes": 1,
        "VpcId": "vpc-3064fb55",
        "NodeType": "dc1.large",
        . . .
    }
}
```

Let's log in to the AWS UI and check out our Redshift cluster:

1  On the AWS dashboard, check that the AWS region in the header at the top right is set to N. Virginia (it is important to select that region for reasons that will become clear in section 10.3).
2  Click Redshift in the Database section.
3  Click the listed cluster, ulp-ch10.

After a few minutes, the status should change from Creating to Available. The details for your cluster should look similar to those shown in figure 10.8.

Our database is alive! But before we can connect to it, we will have to whitelist our current public IP address for access. To do this, we first need to create a security group:

```
$ aws ec2 create-security-group --group-name redshift \
--description
{
    "GroupId": "sg-5b81453c"
}
```

Now let's authorize our IP address, making sure to update the group_id with your own value from the JSON returned by the preceding command:

```
$ group_id=sg-5b81453c
$ public_ip=$(dig +short myip.opendns.com @resolver1.opendns.com)
$ aws ec2 authorize-security-group-ingress --group-id ${group_id} \
--port 5439 --cidr ${public_ip}/32 --protocol tcp --region us-east-1 \
--profile ulp
```

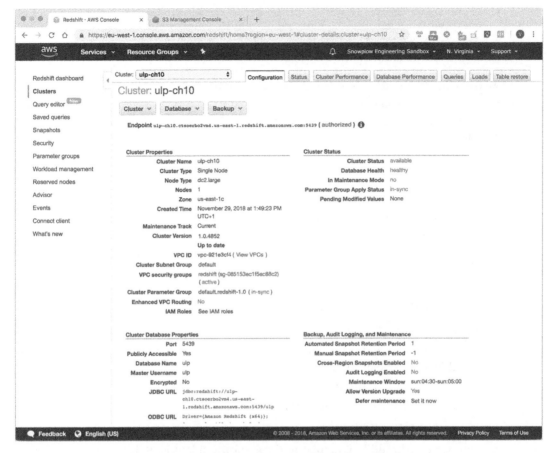

**Figure 10.8   The Configuration tab of the Redshift cluster UI provides all the available metadata for our new cluster, including status information and helpful JDBC and ODBC connection URIs.**

Now we update our cluster to use this security group, again making sure to update the group ID:

```
$ aws redshift modify-cluster --cluster-identifier ulp-ch10 \

{
    "Cluster": {
        "PubliclyAccessible": true,
        "MasterUsername": "ulp",
        "VpcSecurityGroups": [
            {
                "Status": "adding",
                "VpcSecurityGroupId": "sg-5b81453c"
            }
        ],
        ...
```

The returned JSON helpfully shows us the newly added security group near the top. Phew! Now we are ready to connect to our Redshift cluster and check that we can execute SQL queries. We are going to use a standard PostgreSQL client to access our Redshift cluster. If you are working in the book's Vagrant virtual machine, you will find the command-line tool `psql` installed and ready to use. If you have a preferred PostgreSQL GUI, that should work fine too.

First let's check that we can connect to the cluster. Update the first line shown here to point to the Endpoint URI, as shown in the Configuration tab of the Redshift cluster UI:

```
$ host=ulp-ch10.ccxvdpz01xnr.us-east-1.redshift.amazonaws.com
$ export PGPASSWORD=UnifiedLP
$ psql ulp --host ${host} --port 5439 --username ulp
psql (8.4.22, server 8.0.2)
WARNING: psql version 8.4, server version 8.0.
        Some psql features might not work.
SSL connection (cipher: ECDHE-RSA-AES256-SHA, bits: 256)
Type "help" for help.

ulp=#
```

Success! Let's try a simple SQL query to show the first three tables in our database:

```
ulp=# SELECT DISTINCT tablename FROM pg_table_def LIMIT 3;
        tablename
-------------------------
 padb_config_harvest
 pg_aggregate
 pg_aggregate_fnoid_index
(3 rows)
```

Our Redshift cluster is up and running, accessible from our computer's IP address and responding to our SQL queries. We're now ready to start designing the OOPS *event warehouse* to support our analytics-on-read requirements.

### 10.3.3 *Designing an event warehouse*

Remember that at OOPS we have five event types, each tracking a distinct action involving a subset of the five business entities that matter at OOPS: delivery trucks, locations, employees, packages, and customers. We need to store these five event types in a table structure in Amazon Redshift, with maximal flexibility for whatever analytics-on-read we may want to perform in the future.

#### TABLE PER EVENT

How should we store these events in Redshift? One naïve approach would be to define a table in Redshift for each of our five event types. This approach, depicted in figure 10.9, has one obvious issue: to perform any kind of analysis across all event types, we have to use the SQL `UNION` command to join five `SELECT`s together. Imagine

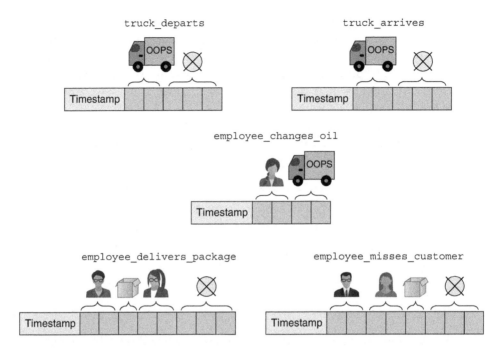

**Figure 10.9   Following the table-per-event approach, we would create five tables for OOPS. Notice how the entities recorded in the OOPS events end up duplicated multiple times across the various event types.**

if we had 30 or 300 event types—even simple counts of events per hour would become extremely painful!

This table-per-event approach has a second issue: our five business entities are duplicated across multiple event tables. For example, the columns for our employee are duplicated in three tables; the columns for a geographical location are in four of our five tables. This is problematic for a few reasons:

- If OOPS decides, say, that all locations should also have a zip code, then we have to upgrade four tables to add the new column.
- If we have a table of employee details in Redshift and we want to JOIN it to the employee ID, we have to write separate JOINs for three separate event tables.
- If we want to analyze a type of entity rather than a type of event (for example, "which locations saw the most event activity on Tuesday?"), then the data we care about is scattered over multiple tables.

If a table per event type doesn't work, what are the alternatives? Broadly, we have two options: a fat table or shredded entities.

### FAT TABLE

A *fat table* refers to a single events table, containing just the following:

- A short tag describing the event (or example, `TRUCK_DEPARTS` or `DRIVER_MISSES_CUSTOMER`)
- A timestamp at which the event took place
- Columns for each of the entities involved in our events

The table is sparsely populated, meaning that columns will be empty if an event does not feature a given entity. Figure 10.10 depicts this approach.

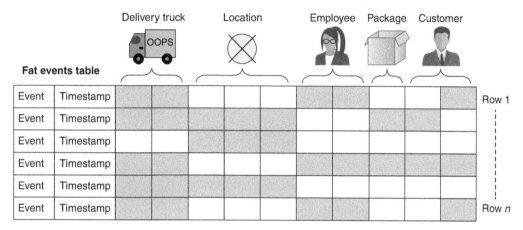

**Figure 10.10** The fat-table approach records all event types in a single table, which has columns for each entity involved in the event. We refer to this table as "sparsely populated" because entity columns will be empty if a given event did not record that entity.

For a simple event model like OOPS's, the fat table can work well: it is straightforward to define, easy to query, and relatively straightforward to maintain. But it starts to show its limits as we try to model more sophisticated events; for example:

- What if we want to start tracking events between two employees (for example, a truck handover)? Do we have to add another set of employee columns to the table?
- How do we support events involving a set of the same entity—for example, when an employee packs *n* items into a package?

Even so, you can go a long way with a fat table of your events in a database such as Redshift. Its simplicity makes it quick to get started and lets you focus on analytics rather than time-consuming schema design. We used the fat-table approach exclusively at Snowplow for about two years before starting to implement another option, described next.

### SHREDDED ENTITIES

Through trial and error, we evolved a third approach at Snowplow, which we refer to as *shredded entities*. We still have a master table of events, but this time it is "thin," containing only the following:

- A short tag describing the event (for example, TRUCK_DEPARTS)
- A timestamp at which the event took place
- An event ID that uniquely identifies the event (UUID4s work well)

Accompanying this master table, we also now have a table per entity; so in the case of OOPS, we would have five additional tables—one each for delivery trucks, locations, employees, packages, and customers. Each entity table contains all the properties for the given entity, but crucially it also includes a column with the event ID of the parent event that this entity belongs to. This relationship allows an analyst to JOIN the relevant entities back to their parent events. Figure 10.11 depicts this approach.

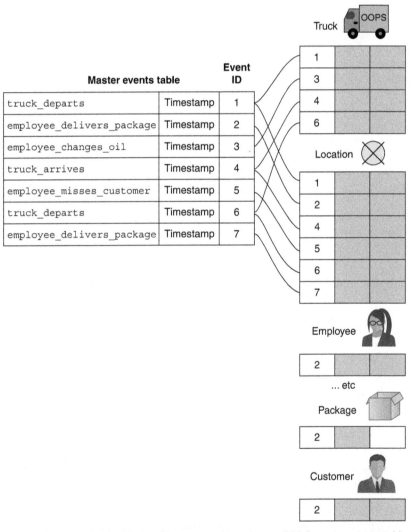

**Figure 10.11  In the shredded-entities approach, we have a "thin" master events table that connects via an event ID to dedicated entity-specific tables. The name comes from the idea that the event has been "shredded" into multiple tables.**

This approach has advantages:

- We can support events that consist of multiple instances of the same entity, such as two employees or *n* items in a package.
- Analyzing entities is super simple: all of the data about a single entity type is available in a single table.
- If our definition of an entity changes (for example, we add a zip code to a location), we have to update only the entity-specific table, not the main table.

Although it is undoubtedly powerful, implementing the shredded-entities approach is complex (and is, in fact, an ongoing project at Snowplow). Our colleagues at OOPS don't have that kind of patience, so for this chapter we are going to design, load, and analyze a fat table instead. Let's get started.

### 10.3.4 *Creating our fat events table*

We already know the design of our fat events table; it is set out in figure 10.10. To deploy this as a table into Amazon Redshift, we will write SQL-flavored data-definition language (DDL); specifically, we need to craft a CREATE TABLE statement for our fat events table.

Writing table definitions by hand is a tedious exercise, especially given that our colleagues at OOPS have already done the work of writing JSON schema files for all five of our event types! Luckily, there is another way: at Snowplow we have open sourced a CLI tool called Schema Guru that can, among other things, autogenerate Redshift table definitions from JSON Schema files. Let's start by downloading the five events' JSON Schema files from the book's GitHub repository:

```
$ git clone https://github.com/alexanderdean/Unified-Log-Processing.git
$ cd Unified-Log-Processing/ch10/10.2 && ls schemas
driver_delivers_package.json   mechanic_changes_oil.json   truck_departs.json
driver_misses_customer.json    truck_arrives.json
```

Now let's install Schema Guru:

```
$ ZIPFILE=schema_guru_0.6.2.zip
$ cd .. && wget http://dl.bintray.com/snowplow/snowplow-generic/${ZIPFILE}
$ unzip ${ZIPFILE}
```

We can now use Schema Guru in *generate DDL mode* against our schemas folder:

```
$ ./schema-guru-0.6.2 ddl –raw-mode ./schemas
File [Unified-Log-Processing/ch10/10.2/./sql/./driver_delivers_package.sql]
 was written successfully!
File [Unified-Log-Processing/ch10/10.2/./sql/./driver_misses_customer.sql]
 was written successfully!
File [Unified-Log-Processing/ch10/10.2/./sql/./mechanic_changes_oil.sql]
 was written successfully!
File [Unified-Log-Processing/ch10/10.2/./sql/./truck_arrives.sql]
 was written successfully!
File [Unified-Log-Processing/ch10/10.2/./sql/./truck_departs.sql]
 was written successfully!
```

After you have run the preceding command, you should have a definition for our fat events table identical to the one in the following listing.

**Listing 10.2   events.sql**

```
CREATE TABLE IF NOT EXISTS events (
    "event"               VARCHAR(23) NOT NULL,
    "timestamp"           TIMESTAMP   NOT NULL,
    "customer.id"         CHAR(36),
    "customer.is_vip"     BOOLEAN,
    "employee.id"         CHAR(36),
    "employee.job_role"   VARCHAR(12),
    "location.elevation"  SMALLINT,
    "location.latitude"   DOUBLE PRECISION,
    "location.longitude"  DOUBLE PRECISION,
    "package.id"          CHAR(36),
    "vehicle.mileage"     INT,
    "vehicle.vin"         CHAR(17)
);
```

To save you the typing, the events table is already available in the book's GitHub repository. We can deploy it into our Redshift cluster like so:

```
ulp=# \i /vagrant/ch10/10.2/sql/events.sql
CREATE TABLE
```

We now have our fat events table in Redshift.

## 10.4   ETL, ELT

Our fat events table is now sitting empty in Amazon Redshift, waiting to be populated with our archive of OOPS events. In this section, we will walk through a simple manual process for loading the OOPS events into our table.

These kinds of processes are often referred to as *ETL*, an old data warehousing acronym for *extract, transform, load*. As MPP databases such as Redshift have grown more popular, the acronym *ELT* has also emerged, meaning that the data is *loaded* into the database *before* further *transformation* is applied.

### 10.4.1   Loading our events

If you have previously worked with databases such as PostgreSQL, SQL Server, or Oracle, you are probably familiar with the COPY statement that loads one or more comma- or tab-separated files into a given table in the database. Figure 10.12 illustrates this approach.

In addition to a regular COPY statement, Amazon Redshift supports something rather more unique: a COPY from JSON statement, which lets us load files of newline-delimited JSON data into a given table.[2] This load process depends on a special file,

---

[2]   The following article describes how to load JSON files into a Redshift table: https://docs.aws.amazon.com/ redshift/latest/dg/copy-usage_notes-copy-from-json.html

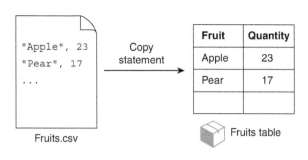

**Figure 10.12** A Redshift COPY statement lets you load one or more comma- or tab-separated flat files into a table in Redshift. The "columns" in the flat files must contain data types that are compatible with the table's corresponding columns.

called a *JSON Paths file*, which provides a mapping from the JSON structure to the table. The format of the JSON Paths file is ingenious. It's a simple JSON array of strings, and each string is a JSON Path expression identifying which JSON property to load into the corresponding column in the table. Figure 10.13 illustrates this process.

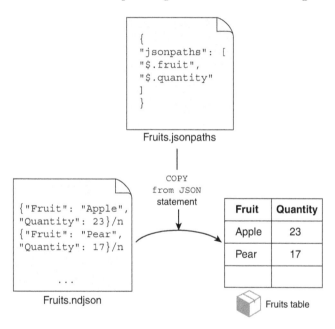

**Figure 10.13** The COPY from JSON statement lets us use a JSON Paths configuration file to load files of newline-separated JSON data into a Redshift table. The array in the JSON Paths file should have an entry for each column in the Redshift table.

If it's helpful, you can imagine that the COPY from JSON statement first creates a temporary CSV or TSV flat file from the JSON data, and then those flat files are COPY'ed into the table.

   In this chapter, we want to create a simple process for loading the OOPS events into Redshift from JSON with as few moving parts as possible. Certainly, we don't want to have to write any (non-SQL) code if we can avoid it. Happily, Redshift's COPY from JSON statement should let us load our OOPS event archive directly into our fat events table. A further stroke of luck: we can use a single JSON Paths file to load all five of our event types! This is because OOPS's five event types have a predictable and shared structure (the five entities again), and because a COPY from JSON can tolerate specified JSON Paths not being found in a given JSON (it will simply set those columns to null).

We are now ready to write our JSON Paths file. Although Schema Guru can generate these files automatically, it's relatively simple to do this manually: we just need to use the correct syntax for the JSON Paths and ensure that we have an entry in the array for each column in our fat events table. The following listing contains our populated JSON Paths file.

**Listing 10.3   events.jsonpaths**

```
{
  "jsonpaths": [
    "$.event",                  The standard fields
    "$.timestamp",              shared by all events
    "$.customer.id",        ◁─────
    "$.customer.isVip",
    "$.employee.id",            JSON Paths statements to
    "$.employee.jobRole",       extract each property for
    "$.location.elevation",     each entity into its own
    "$.location.latitude",      table column
    "$.location.longitude",
    "$.package.id",
    "$.vehicle.mileage",
    "$.vehicle.vin"
  ]
}
```

A Redshift COPY from JSON statement requires the JSON Paths file to be available on Amazon S3, so we have uploaded the events.jsonpaths file to Amazon S3 at the following path:

```
s3://ulp-assets-2019/ch10/jsonpaths/event.jsonpaths
```

We are now ready to load all of the OOPS event archive into Redshift! We will do this as a one-time action, although you could imagine that if our analytics-on-read are fruitful, our OOPS colleagues could configure this load process to occur regularly—perhaps nightly or hourly. Reopen your psql connection if it has closed:

```
$ psql ulp --host ${host} --port 5439 --username ulp
...
ulp=#
```

Now execute your COPY from JSON statement, making sure to update your AWS access key ID and secret access key (the XXXs) accordingly:

```
ulp=# COPY events FROM 's3://ulp-assets-2019/ch10/jsonpaths/data/' \
 CREDENTIALS 'aws_access_key_id=XXX;aws_secret_access_key=XXX' JSON \
 's3://ulp-assets-2019/ch10/jsonpaths/event.jsonpaths' \
REGION 'us-east-1' TIMEFORMAT 'auto';
INFO:  Load into table 'events' completed, 140 record(s) loaded successfully.
COPY
```

Here are some notes on the unfamiliar parts of the syntax:

- The AWS CREDENTIALS will be used to access the data in S3 and the JSON Paths file.
- The REGION parameter specifies the region in which the data and JSON Paths file are located. If you get an error similar to S3ServiceException: The bucket you are attempting to access must be addressed using the specified endpoint, it means that your Redshift cluster and your S3 bucket are in different regions. Both must be in the same region for the COPY command to succeed.
- TIMEFORMAT 'auto' allows the COPY command to autodetect the timestamp format used for a given input field.

Let's perform a simple query now to get an overview of the events we have loaded:

```
ulp=# SELECT event, COUNT(*) FROM events GROUP BY 1 ORDER BY 2 desc;
          event          | count
-------------------------+-------
 TRUCK_DEPARTS           |    52
 TRUCK_ARRIVES           |    52
 MECHANIC_CHANGES_OIL    |    19
 DRIVER_DELIVERS_PACKAGE |     5
 DRIVER_MISSES_CUSTOMER  |     2
(5 rows)
```

Great—we have loaded some OOPS delivery events into Redshift!

### 10.4.2 *Dimension widening*

We now have our event archive loaded into Redshift—our OOPS teammates will be pleased! But let's take a look at an individual event:

```
\x on
Expanded display is on.
ulp=# SELECT * FROM events WHERE event='DRIVER_MISSES_CUSTOMER' LIMIT 1;
-[ RECORD 1 ]------+-------------------------------------
event              | DRIVER_MISSES_CUSTOMER
timestamp          | 2018-11-11 12:27:00
customer.id        | 4594f1a1-a7a2-4718-bfca-6e51e73cc3e7
customer.is_vip    | f
employee.id        | 54997a47-252d-499f-a54e-1522ac49fa48
employee.job_role  | JNR_DRIVER
location.elevation | 102
location.latitude  | 51.4972997
location.longitude | -0.0955459
package.id         | 14a714cf-5a89-417e-9c00-f2dba0d1844d
vehicle.mileage    |
vehicle.vin        |
```

Doesn't it look a bit, well, uninformative? The event certainly has lots of identifiers, but few interesting data points: a VIN uniquely identifies an OOP delivery truck, but it doesn't tell us how old the truck is, or the truck's make or model. It would be ideal to look up these entity identifiers in a reference database, and then add this extra entity data to our events. In analytics speak, this is called *dimension widening*, because we are taking one of the dimensions in our event data and widening it with additional data points.

Imagine that we ask our colleagues about this, and one of them shares with us a Redshift-compatible SQL file of OOPS reference data. The SQL file creates and populates four tables covering same of the entities involved in our events: vehicles, employees, customers, and packages. OOPS has no reference data for locations, but no matter. Longitude, latitude, and elevation give us ample information about each location. Each of the four entity tables contains an ID that can be used to join a given row with a corresponding entity in our events table. Figure 10.14 illustrates these relationships.

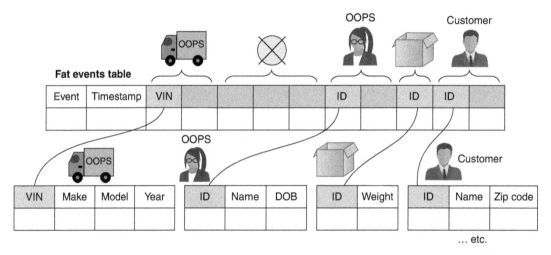

**Figure 10.14   We can use the entity identifiers in our fat-events table to join our event back to per-entity reference tables. This dimension widening gives us much richer events to analyze.**

The SQL file for our four reference tables and their rows is shown in listing 10.4. This is deliberately a hugely abbreviated reference dataset. In reality, a company such as OOPS would have significant volumes of reference data, and synchronizing that data into the event warehouse to support analytics-on-read would be a significant and ongoing ETL project itself.

**Listing 10.4   reference.sql**

```
CREATE TABLE vehicles(
  vin CHAR(17) NOT NULL,
  make VARCHAR(32) NOT NULL,
  model VARCHAR(32) NOT NULL,
  year SMALLINT);
INSERT INTO vehicles VALUES
  ('1HGCM82633A004352', 'Ford', 'Transit', 2005),
```

```
  ('JH4TB2H26CC000000', 'VW', 'Caddy', 2010),
  ('19UYA31581L000000', 'GMC', 'Savana', 2011);

CREATE TABLE employees(
  id CHAR(36) NOT NULL,
  name VARCHAR(32) NOT NULL,
  dob DATE NOT NULL);
INSERT INTO employees VALUES
  ('f2caa6a0-2ce8-49d6-b793-b987f13cfad9', 'Amanda', '1992-01-08'),
  ('f6381390-32be-44d5-9f9b-e05ba810c1b7', 'Rohan', '1983-05-17'),
  ('3b99f162-6a36-49a4-ba2a-375e8a170928', 'Louise', '1978-11-25'),
  ('54997a47-252d-499f-a54e-1522ac49fa48', 'Carlos', '1985-10-27'),
  ('c4b843f2-0ef6-4666-8f8d-91ac2e366571', 'Andreas', '1994-03-13');

CREATE TABLE packages(
  id CHAR(36) NOT NULL,
  weight INT NOT NULL);
INSERT INTO packages VALUES
  ('c09e4ee4-52a7-4cdb-bfbf-6025b60a9144', 564),
  ('ec99793d-94e7-455f-8787-1f8ebd76ef61', 1300),
  ('14a714cf-5a89-417e-9c00-f2dba0d1844d', 894),
  ('834bc3e0-595f-4a6f-a827-5580f3d346f7', 3200),
  ('79fee326-aaeb-4cc6-aa4f-f2f98f443271', 2367);

CREATE TABLE customers(
  id CHAR(36) NOT NULL,
  name VARCHAR(32) NOT NULL,
  zip_code VARCHAR(10) NOT NULL);
INSERT INTO customers VALUES
  ('b39a2b30-049b-436a-a45d-46d290df65d3', 'Karl', '99501'),
  ('4594f1a1-a7a2-4718-bfca-6e51e73cc3e7', 'Maria', '72217-2517'),
  ('b1e5d874-963b-4992-a232-4679438261ab', 'Amit', '90089');
```

To save you the typing, this reference data is again available in the book's GitHub repository. You can run it against your Redshift cluster like so:

```
ulp=# \i /vagrant/ch10/10.4/sql/reference.sql
CREATE TABLE
INSERT 0 3
CREATE TABLE
INSERT 0 5
CREATE TABLE
INSERT 0 5
CREATE TABLE
INSERT 0 3
```

Let's get a feel for the reference data with a simple LEFT JOIN of the vehicles table back onto the events table:

```
ulp=# SELECT e.event, e.timestamp, e."vehicle.vin", v.* FROM events e \
 LEFT JOIN vehicles v ON e."vehicle.vin" = v.vin LIMIT 1;
-[ RECORD 1 ]--------------------
event      | TRUCK_ARRIVES
timestamp  | 2018-11-01 03:37:00
```

```
vehicle.vin | 1HGCM82633A004352
vin         | 1HGCM82633A004352
make        | Ford
model       | Transit
year        | 2005
```

The event types that involve a vehicle entity will have all of the fields from the vehicles table (v.*) populated. This is a good start, but we don't want to have to manually construct these JOINs for every query. Instead, let's create a single Redshift view that joins all of our reference tables back to the fat events table, to create an even fatter table. This view is shown in the following listing.

**Listing 10.5   widened.sql**

```
CREATE VIEW widened AS
  SELECT
    ev."event"                AS "event",
    ev."timestamp"            AS "timestamp",
    ev."customer.id"          AS "customer.id",
    ev."customer.is_vip"      AS "customer.is_vip",
    c."name"                  AS "customer.name",
    c."zip_code"              AS "customer.zip_code",
    ev."employee.id"          AS "employee.id",
    ev."employee.job_role"    AS "employee.job_role",
    e."name"                  AS "employee.name",
    e."dob"                   AS "employee.dob",
    ev."location.latitude"    AS "location.latitude",
    ev."location.longitude"   AS "location.longitude",
    ev."location.elevation"   AS "location.elevation",
    ev."package.id"           AS "package.id",
    p."weight"                AS "package.weight",
    ev."vehicle.vin"          AS "vehicle.vin",
    ev."vehicle.mileage"      AS "vehicle.mileage",
    v."make"                  AS "vehicle.make",
    v."model"                 AS "vehicle.model",
    v."year"                  AS "vehicle.year"
  FROM events ev
    LEFT JOIN vehicles v  ON ev."vehicle.vin" = v.vin
    LEFT JOIN employees e ON ev."employee.id" = e.id
    LEFT JOIN packages p  ON ev."package.id" = p.id
    LEFT JOIN customers c ON ev."customer.id" = c.id;
```

Again, this is available from the GitHub repository, so you can run it against your Redshift cluster like so:

```
ulp=# \i /vagrant/ch10/10.4/sql/widened.sql
CREATE VIEW
```

Let's now get an event back from the view:

```
ulp=# SELECT * FROM widened WHERE event='DRIVER_MISSES_CUSTOMER' LIMIT 1;
-[ RECORD 1 ]------+------------------------------------
event              | DRIVER_MISSES_CUSTOMER
```

```
timestamp            | 2018-11-11 12:27:00
customer.id          | 4594f1a1-a7a2-4718-bfca-6e51e73cc3e7
customer.is_vip      | f
customer.name        | Maria
customer.zip_code    | 72217-2517
employee.id          | 54997a47-252d-499f-a54e-1522ac49fa48
employee.job_role    | JNR_DRIVER
employee.name        | Carlos
employee.dob         | 1985-10-27
location.latitude    | 51.4972997
location.longitude   | -0.0955459
location.elevation   | 102
package.id           | 14a714cf-5a89-417e-9c00-f2dba0d1844d
package.weight       | 894
vehicle.vin          |
vehicle.mileage      |
vehicle.make         |
vehicle.model        |
vehicle.year         |
```

That's much better! Our dimension-widened event now has plenty of interesting data points in it, courtesy of our new view. Note that views in Redshift are not *physically materialized*, meaning that the view's underlying query is executed every time the view is referenced in a query. Still, we can approximate a materialized view by simply loading the view's contents into a new table:

```
ulp=# CREATE TABLE events_w AS SELECT * FROM widened;
SELECT
```

Running queries against our newly created events_w table will be much quicker than going back to the widened view each time.

### 10.4.3 *A detour on data volatility*

At this point, you might be wondering why some of our data points are embedded in the event (such as the delivery truck's mileage), whereas other data points (such as the delivery truck's year of registration) are joined to the event later, in Redshift. You could say that the data points embedded inside the event were *early*, or *eagerly* joined, to the event, whereas the data points joined only in Redshift were *late*, or *lazily* joined. When our OOPS coworkers made these decisions, long before our arrival on the team, what drove them?

The answer comes down to the *volatility*, or changeability, of the individual data point. We can broadly divide data points into three levels of volatility:

- *Stable data points*—For example, the delivery truck's year of registration or the delivery driver's date of birth
- *Slowly or infrequently changing data points*—For example, the customer's VIP status, the delivery driver's current job role, or the customer's name
- *Volatile data points*—For example, the delivery truck's current mileage

The volatility of a given data point is not set in stone: a customer's surname might change after marriage, whereas the truck's mileage won't change while it is in the garage having its oil changed. But the *expected volatility* of a given data point gives us some guidance on how we should track it: volatile data points should be eagerly joined in our event tracking, assuming they are available. Unchanging data points can be lazily joined later, in our unified log. For slowly changing data points, we need to be pragmatic; we may choose to eagerly join some and lazily join others. This is visualized in figure 10.15.

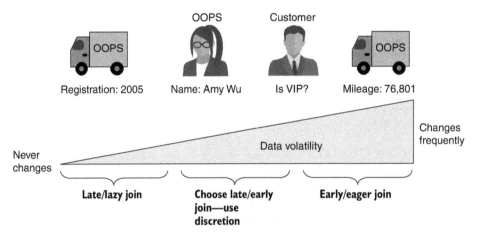

**Figure 10.15   The volatility of a given data point influences whether we should attach that data point to our events "early" or "late."**

## 10.5   *Finally, some analysis*

By this point, you will have noticed that this chapter isn't primarily about performing analytics-on-read. Rather, we have focused on putting the processes and tooling in place to support future analytics-on-read efforts, whether performed by you or someone else (perhaps in your BI team). Still, we can't leave this chapter without putting Redshift and our SQL skills through their paces.

### 10.5.1   *Analysis 1: Who does the most oil changes?*

From our reference data, we know that OOPS has two mechanics, but are they both doing their fair share of oil changes?

```
ulp=# \x off
Expanded display is off.
ulp=# SELECT "employee.id", "employee.name", COUNT(*) FROM events_w
 WHERE event='MECHANIC_CHANGES_OIL' GROUP BY 1, 2;

            employee.id              | employee.name | count
-------------------------------------+---------------+-------
 f6381390-32be-44d5-9f9b-e05ba810c1b7 | Rohan         |    15
 f2caa6a0-2ce8-49d6-b793-b987f13cfad9 | Amanda        |     4
(2 rows)
```

Interesting! Rohan is doing around three times the number of oil changes as Amanda. Perhaps this has something to do with seniority. Let's repeat the query, but this time including the mechanic's job at the time of the oil change:

```
ulp=# SELECT "employee.id", "employee.name" AS name, "employee.job_role"
 AS job, COUNT(*) FROM events_w WHERE event='MECHANIC_CHANGES_OIL'

              employee.id               | name   |     job      | count
----------------------------------------+--------+--------------+-------
 f6381390-32be-44d5-9f9b-e05ba810c1b7   | Rohan  | SNR_MECHANIC |   6
 f2caa6a0-2ce8-49d6-b793-b987f13cfad9   | Amanda | SNR_MECHANIC |   4
 f6381390-32be-44d5-9f9b-e05ba810c1b7   | Rohan  | JNR_MECHANIC |   9
(3 rows)
```

We can now see that Rohan received a promotion to senior mechanic partway through the event stream. Perhaps because of all the oil changes he has been doing! When exactly did Rohan get his promotion? Unfortunately, the OOPS HR system isn't wired into our unified log, but we can come up with an approximate date for his promotion:

```
ulp=# SELECT MIN(timestamp) AS range FROM events_w WHERE
 "employee.name" = 'Rohan' AND "employee.job_role" = 'SNR_MECHANIC'
 UNION SELECT MAX(timestamp) AS range FROM events_w WHERE
 "employee.name" = 'Rohan' AND "employee.job_role" = 'JNR_MECHANIC';
       range
---------------------
 2018-12-05 01:11:00
 2018-12-05 10:58:00
(2 rows)
```

Rohan was promoted sometime on the morning of December 5, 2018.

### 10.5.2 Analysis 2: Who is our most unreliable customer?

If a driver misses a customer, the package has to be taken back to the depot and another attempt has to be made to deliver the package. Are some OOPS customers less reliable than others? Are they consistently out when they promise to be in? Let's check:

```
ulp=# SELECT "customer.name", SUM(CASE WHEN event LIKE '%_DELIVERS_%'
 THEN 1 ELSE 0 END) AS "delivers", SUM(CASE WHEN event LIKE '%_MISSES_%'
 THEN 1 ELSE 0 END) AS "misses" FROM events_w WHERE event LIKE 'DRIVER_%'
 GROUP BY "customer.name";

 customer.name | delivers | misses
---------------+----------+--------
 Karl          |        2 |      0
 Maria         |        2 |      2
 Amit          |        1 |      0
(3 rows)
```

This gives us the answer in a rough form, but we can make this a little more precise with a sub-select:

```
ulp=# SELECT "customer.name", 100 * misses/count AS "miss_pct" FROM
 (SELECT "customer.name", COUNT(*) AS count, SUM(CASE WHEN event LIKE
 '%_MISSES_%' THEN 1 ELSE 0 END) AS "misses" FROM events_w WHERE event
LIKE 'DRIVER_%' GROUP BY "customer.name");
 customer.name | miss_pct
---------------+----------
 Karl          |        0
 Maria         |       50
 Amit          |        0
(3 rows)
```

Maria has a 50% miss rate! I do hope for OOPS's sake that she's not a VIP:

```
ulp=# SELECT COUNT(*) FROM events_w WHERE "customer.name" = 'Maria'
 AND "customer.is_vip" IS true;
 count
-------
     0
(1 row)
```

No, she isn't a VIP. Let's see which drivers have been unluckiest delivering to Maria:

```
ulp=# SELECT "timestamp", "employee.name", "employee.job_role" FROM
 events_w WHERE event LIKE 'DRIVER_MISSES_CUSTOMER';
      timestamp      | employee.name | employee.job_role
---------------------+---------------+-------------------
 2018-01-11 12:27:00 | Carlos        | JNR_DRIVER
 2018-01-10 21:53:00 | Andreas       | JNR_DRIVER
(2 rows)
```

So, it looks like Andreas and Carlos have both missed Maria, once each. And this concludes our brief foray into analytics-on-read. If your interest has been piqued, we encourage you to continue exploring the OOPS event dataset. You can find the Redshift SQL reference here, along with plenty of examples that can be adapted to the OOPS events fairly easily: https://docs.aws.amazon.com/redshift/latest/dg/cm_chap_SQLCommandRef.html.

In the next chapter, we will explore analytics-on-write.

## Summary

- We can divide event stream analytics into analytics-on-read and analytics-on-write.
- Analytics-on-read means "storing first, asking questions later." We aim to store our events in one or more storage targets in a format that facilitates performing a wide range of analysis later.
- Analytics-on-write involves defining our analysis ahead of time and performing it in real time as the events stream in. This is a great fit for dashboards, operational reporting, and other low-latency use cases.
- Amazon Redshift is a hosted columnar database that scales horizontally and offers full Postgres-like SQL.

- We can model our event stream in a columnar database in various ways: one table per event type, a single fat events table, or a master table with child tables, one per entity type.

- We can load our events stored in flat files as JSON data into Redshift by using `COPY from JSON` and a manifest file that contains JSON Paths statements. Schema Guru can help us automatically generate the Redshift table definition from the event JSON schema data.

- Once our events are loaded, we can join those events back to reference tables in Redshift based on shared entity IDs; this is called *dimension widening*.

- With the events loaded into Redshift and widened, we or our colleagues in BI can perform a wide variety of analyses using SQL.

# 11

# *Analytics-on-write*

**This chapter covers**

- Simple algorithms for analytics-on-write on event streams
- Modeling operational reporting as a DynamoDB table
- Writing an AWS Lambda function for analytics-on-write
- Deploying and testing an AWS Lambda function

In the previous chapter, we implemented a simple analytics-on-read strategy for OOPS, our fictitious package-delivery company, using Amazon Redshift. The focus was on storing our event stream in Redshift in such a way as to support as many analyses as possible "after the fact." We modeled a fat event table, widened it further with dimension lookups for key entities including drivers and trucks, and then tried out a few analyses on the data in SQL.

For the purposes of this chapter, we will assume that some time has passed at OOPS, during which the BI team has grown comfortable with writing SQL queries against the OOPS event stream as stored in Redshift. Meanwhile, rumblings are coming from various stakeholders at OOPS who want to see analyses that are not well suited to Redshift. For example, they are interested in the following:

- *Low-latency operational reporting*—This must be fed from the incoming event streams in as close to real time as possible.
- *Dashboards to support thousands of simultaneous users*—For example, a parcel tracker on the website for OOPS customers.

In this chapter, we will explore techniques for delivering these kinds of analytics, which broadly fall under the term *analytics-on-write*. Analytics-on-write has an up-front cost: we have to decide on the analysis we want to perform ahead of time and put this analysis live on our event stream. In return for this constraint, we get some benefits: our queries are low latency, can serve many simultaneous users, and are simple to operate.

To implement analytics-on-write for OOPS, we are going to need a database for storing our aggregates, and a stream processing framework to turn our events into aggregates. For the database, we will use Amazon DynamoDB, which is a hosted, highly scalable key-value store with a relatively simple query API. For the stream processing piece, we are going to try out AWS Lambda, which is an innovative platform for single-event processing, again fully hosted by Amazon Web Services.

Let's get started!

## 11.1 Back to OOPS

For this chapter, we will be returning to OOPS, the scene of our analytics-on-read victory, to implement analytics-on-write. Before we can get started, we need to get Kinesis set up and find out exactly what analytics our OOPS bosses are expecting.

### 11.1.1 Kinesis setup

In the preceding chapter, we interacted with the OOPS unified log through an archive of events stored in Amazon S3. This archive contained the five types of OOPS events, all stored in JSON, as recapped in figure 11.1. This event archive suited our analytics-on-read requirements fine, but for analytics-on-write, we need something much "fresher": we need access to the OOPS event stream as it flows through Amazon Kinesis in near real-time.

Our OOPS colleagues have not yet given us access to the "firehose" of live events in Kinesis, but they have shared a Python script that can generate valid OOPS events and write them directly to a Kinesis stream for testing. You can find this script in the GitHub repository:

```
ch11/11.1/generate.py
```

Before we can use this script, we must first set up a new Kinesis stream to write the events to. As in chapter 4, we can do this by using the AWS CLI tools:

```
$ aws kinesis create-stream --stream-name oops-events \
  --shard-count 2 --region=us-east-1 --profile=ulp
```

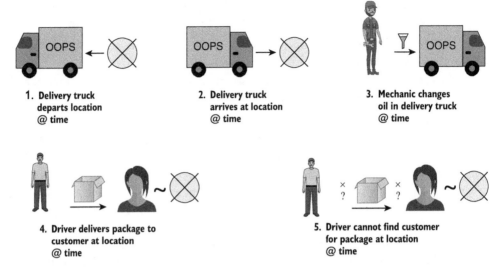

**Figure 11.1   The five types of events generated at OOPS are unchanged since their introduction in the previous chapter.**

That command doesn't give any output, so let's follow up by describing the stream:

```
$ aws kinesis describe-stream --stream-name oops-events \
  --region=us-east-1 --profile=ulp
{
    "StreamDescription": {
        "StreamStatus": "ACTIVE",
        "StreamName": "oops-events",
        "StreamARN": "arn:aws:kinesis:us-east-1:719197435995:stream/
  oops-events",
...
```

We are going to need the stream's ARN later, so let's add it into an environment variable:

```
$ stream_arn=arn:aws:kinesis:us-east-1:719197435995:stream/oops-events
```

Now let's try out the event generator:

```
$ /vagrant/ch11/11.1/generate.py
Wrote TruckArrivesEvent with timestamp 2018-01-01 00:14:00
Wrote DriverMissesCustomer with timestamp 2018-01-01 02:14:00
Wrote DriverDeliversPackage with timestamp 2018-01-01 04:03:00
```

Great—we can successfully write events to our Kinesis stream. Press Ctrl-C to cancel the generator.

### 11.1.2 Requirements gathering

With analytics-on-read, our focus was on devising the most flexible way of storing our event stream in Redshift, so that we could support as great a variety of *post facto* analyses as possible. The structure of our event storage was optimized for future flexibility, rather than being tied to any specific analysis that OOPS might want to perform later.

With analytics-on-write, the opposite is true: we must understand *exactly* what analysis OOPS wants to see, so that we can build that analysis to run in near real-time on our Kinesis event stream. And, ideally, we would get this right the first time; remember that a Kinesis stream stores only the last 24 hours of events (configurable up to one week), after which the events auto-expire. If there are any mistakes in our analytics-on-write processing, at best we will be able to rerun for only the last 24 hours (or one week). This is visualized in figure 11.2.

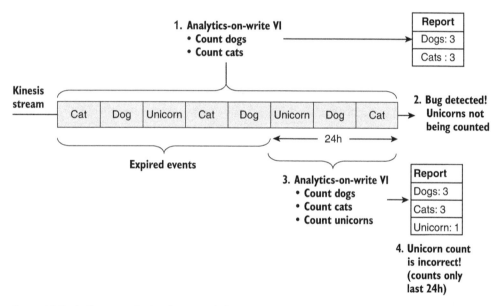

**Figure 11.2 In the case of a bug in our analytics-on-write implementation, we have to fix the bug and redeploy the analytics against our event stream. At best, we can recover the last 24 hours of missing data, as depicted here. At worst, our output is corrupted, and we have to restart our analytics from scratch.**

This means that we need to sit down with the bigwigs at OOPS and find out exactly what near-real-time analytics they want to see. When we do this, we learn that the priority is around operational reporting about the delivery trucks. In particular, OOPS wants a near-real-time feed that tells them the following:

- The location of each delivery truck
- The number of miles each delivery truck has driven since its last oil change

This sounds like a straightforward analysis to generate. We can sketch out a simple table structure that holds all of the relevant data, as shown in table 11.1.

**Table 11.1   The status of OOPS's delivery trucks**

| Truck VIN | Latitude | Longitude | Miles since oil change |
|---|---|---|---|
| 1HGCM82633A004352 | 51.5208046 | -0.1592323 | 35 |
| JH4TB2H26CC000000 | 51.4972997 | -0.0955459 | 167 |
| 19UYA31581L000000 | 51.4704679 | -0.1176902 | 78 |

Now we have our five event types flowing into Kinesis, and we understand the analysis that our OOPS coworkers are looking for. In the next section, let's define the analytics-on-write algorithm that will connect the dots.

### 11.1.3 *Our analytics-on-write algorithm*

We need to create an algorithm that will use the OOPS event stream to populate our four-column table and keep it up-to-date. Broadly, this algorithm needs to do the following:

1  Read each event from Kinesis.
2  If the event involves a delivery truck, use its current mileage to update the count of miles since the last oil change.
3  If the event is an oil change, reset the count of miles since the last oil change.
4  If the event associates a delivery truck with a specific location, update the table row for the given truck with the event's latitude and longitude.

Our approach will have similarities to the *stateful event processing* that we implemented in chapter 5. In this case, we are not interested in processing multiple events at a time, but we do need to make sure that we are always updating our table with a truck's most recent location. There is a risk that if our algorithm ingests an older event after a newer event, it could incorrectly overwrite the "latest" latitude and longitude with the older one. This danger is visualized in figure 11.3.

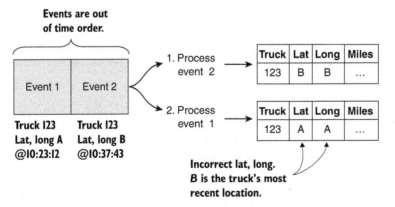

**Figure 11.3   If an older event is processed after a newer event, a naïve algorithm could accidentally update our truck's "current" location with the older event's location.**

We can prevent this by storing an additional piece of state in our table, as shown in table 11.2. The new column, *Location timestamp*, records the timestamp of the event from which we have taken each truck's current latitude and longitude. Now when we process an event from Kinesis containing a truck's location, we check this event's timestamp against the existing Location timestamp in our table. Only if the new event's timestamp "beats" the existing timestamp (it's more recent) do we update our truck's latitude and longitude.

**Table 11.2  Adding the *Location timestamp* column to our table**

| Truck VIN | Latitude | Longitude | Location timestamp | Miles since oil change |
|-----------|----------|-----------|--------------------|------------------------|
| 1HGCM8... | 51.5208046 | -0.1592323 | 2018-08-02T21:50:49Z | 35 |
| JH4TB2... | 51.4972997 | -0.0955459 | 2018-08-01T22:46:12Z | 167 |
| 19UYA3... | 51.4704679 | -0.1176902 | 2018-08-02T18:14:45Z | 78 |

This leaves only our last metric, *Miles since oil change*. This one is slightly more complicated to calculate. The trick is to realize that we should not be storing this metric in our table. Instead, we should be storing the two inputs that go into calculating this metric:

- Current (latest) mileage
- Mileage at the time of the last oil change

With these two metrics stored in our table, we can calculate *Miles since oil change* whenever we need it, like so:

```
miles since oil change = current mileage - mileage at last oil change
```

For the cost of a simple calculation at serving time, we have a much easier-to-maintain table, as you will see in the next section. Table 11.3 shows the final version of our table structure, with the new *Mileage* and *Mileage at oil change* columns.

**Table 11.3  Replacing our *Miles since oil change* metric with *Mileage* and *Mileage at oil change***

| Truck VIN | Latitude | Longitude | Location timestamp | Mileage | Mileage at oil change |
|-----------|----------|-----------|--------------------|---------|-----------------------|
| 1HGCM8... | 51.5... | -0.15... | 2018-08-02T21:50:49Z | 12453 | 12418 |
| JH4TB2... | 51.4... | -0.09... | 2018-08-01T22:46:12Z | 19090 | 18923 |
| 19UYA3... | 51.4... | -0.11... | 2018-08-02T18:14:45Z | 8407 | 8329 |

Remember that for this table to be accurate, we always want to be recording the truck's most recent mileage, and the truck's mileage at its last-known oil change. Again, how do we handle the situation where an *older* event arrives after a *more recent* event? This time, we have something in our favor, which is that a truck's mileage *monotonically increases* over time, as per figure 11.4. Given two mileages for the same truck,

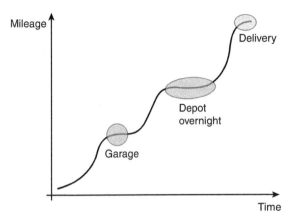

**Figure 11.4** **The mileage recorded on a given truck's odometer monotonically increases over time. We can see periods where the mileage is flat, but it never decreases as time progresses.**

the higher mileage is always the more recent one, and we can use this rule to unambiguously discard the mileages of older events.

Putting all this together, we can now fully specify our analytics-on-write algorithm in pseudocode. First of all, we know that all of our analytics relate to trucks, so we should start with a filter that excludes any events that don't include a vehicle entity:

```
let e = event to process
if e.event not in ("TRUCK_ARRIVES", "TRUCK_DEPARTS",
  "MECHANIC_CHANGES_OIL") then
  skip event
end if
```

Now let's process our vehicle's mileage:

```
let vin = e.vehicle.vin
let event_mi = e.vehicle.mileage
let current_mi = table[vin].mileage
if current_mi is null or current_mi < event_mi then
    set table[vin].mileage = event_mi
end if
```

Most of this pseudocode should be self-explanatory. The trickiest part is the `table[vin] .mileage` syntax, which references the value of a column for a given truck in our table. The important thing to understand here is that we update a truck's mileage in our table only if the event being processed reports a *higher* mileage than the one already in our table for this truck.

Let's move on our truck's location, which is captured only in events relating to a truck arriving or departing:

```
if e.event == "TRUCK_ARRIVES" or "TRUCK_DEPARTS" then
    let ts = e.timestamp
    if ts > table[vin].location_timestamp then
      set table[vin].location_timestamp = ts
      set table[vin].latitude = e.location.latitude
```

```
        set table[vin].longitude = e.location.longitude
    end if
end if
```

The logic here is fairly simple. Again, we have an `if` statement that makes updating the location conditional on the event's timestamp being newer than the one currently found in our table. This is to ensure that we don't accidentally overwrite our table with a stale location.

Finally, we need to handle any oil-change events:

```
if e.event == "MECHANIC_CHANGES_OIL" then
    let current_maoc = table[vin].mileage_at_oil_change
    if current_maoc is null or event_mi > current_maoc then
        set table[vin].mileage_at_oil_change = event_mi
    end if
end if
```

Again, we have a "guard" in the form of an `if` statement, which makes sure that we update the *Mileage at oil change* for this truck only if this oil-change event is newer than any event previously fed into the table. Putting it all together, we can see that this guard-heavy approach is safe even in the case of out-of-order events, such as when an oil-change event is processed *after* a *subsequent* truck-departs event. This is shown in figure 11.5.

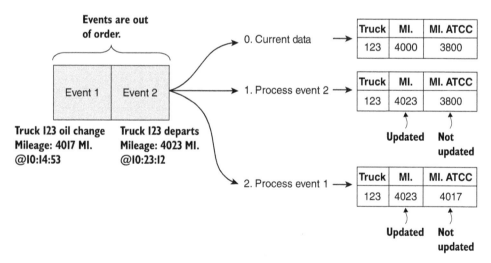

**Figure 11.5** The `if` statements in our analytics-on-write algorithm protect us from out-of-order events inadvertently overwriting current metrics in our table with stale metrics.

If OOPS were a real company, it would be crucial to get this algorithm reviewed and signed off before implementing the stream processing job. With analytics-on-read, if we make a mistake in a query, we just cancel it and run the revised query. With analytics-on-write, a mistake in our algorithm could mean having to start over from scratch. Happily, in our case, we can move swiftly onto the implementation!

## 11.2   *Building our Lambda function*

Enough theory and pseudocode—in this section, we will implement our analytics-on-write by using AWS Lambda, populating our delivery-truck statuses to a table in DynamoDB. Let's get started.

### 11.2.1   *Setting up DynamoDB*

We are going to write a stream processing job that will read OOPS events, cross-check them against existing values in our table, and then update rows in that table accordingly. Figure 11.6 shows this flow.

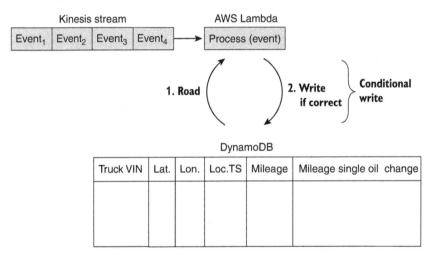

**Figure 11.6   Our AWS Lambda function will read individual events from OOPS, check whether our table in DynamoDB can be updated with the event's values, and update the row in DynamoDB if so. This approach uses a DynamoDB feature called conditional writes.**

We already have a good idea of the algorithm we are going to implement, but rather than jumping into that, let's work backward from the table we need to populate. This table is going to live in Amazon's DynamoDB, a highly scalable hosted key-value store that is a great fit for analytics-on-write on AWS. We can create our table in DynamoDB by using the AWS CLI tools like so:

```
$ aws dynamodb create-table --table-name oops-trucks \
  --provisioned-throughput ReadCapacityUnits=5,WriteCapacityUnits=5 \
  --attribute-definitions AttributeName=vin,AttributeType=S \
  --key-schema AttributeName=vin,KeyType=HASH --profile=ulp \
  --region=us-east-1
{
    "TableDescription": {
```

```
        "TableArn": "arn:aws:dynamodb:us-east-1:719197435995:table/
⇒ oops-trucks",
...
```

When creating a DynamoDB table, we have to specify only a few properties up-front:

- The `--provisioned-throughput` tells AWS how much throughput we want to reserve for reading and writing data from the table. For the purposes of this chapter, low values are fine for both.
- The `--attribute-definitions` defines an attribute for the vehicle identification number, which is a `string` called `vin`.
- The `--key-schema` specifies that the `vin` attribute will be our table's primary key.

Our table has now been created. We can now move on to creating our AWS Lambda function.

### 11.2.2 *Introduction to AWS Lambda*

The central idea of AWS Lambda is that developers should be writing *functions*, not *servers*. With Lambda, we write self-contained functions to process events, and then we publish those functions to Lambda to run. We don't worry about developing, deploying, or managing servers. Instead, Lambda takes care of scaling out our functions to meet the incoming event volumes. Of course, Lambda functions run on servers under the hood, but those servers are abstracted away from us.

At the time of writing, we can write our Lambda functions to run either on Node.js or on the JVM. In both cases, the function has a similar API, in pseudocode:

```
def recordHandler(events: List[Event])
```

This function signature has some interesting features:

- The function operates on a list or array of events, rather than a single one. Under the hood, Lambda is collecting a *microbatch* of events before invoking the function.
- The function does not return anything. It exists only to produce *side effects*, such as writing to DynamoDB or creating new Kinesis events.

Figure 11.7 illustrates these two features of a Lambda function.

Does this function API look familiar? Remember back in chapter 5 when we wrote event-processing jobs using Apache Samza, the API looked like this:

```
public void process(IncomingMessageEnvelope envelope,
    MessageCollector collector, TaskCoordinator coordinator)
```

In both cases, we are implementing a side-effecting function that is invoked for incoming events—or in Samza's case, a single event. And as a Lambda function is run on AWS Lambda, so our Samza jobs were run on Apache YARN: we can say that Lambda and YARN are the execution environments for our functions. The similarities

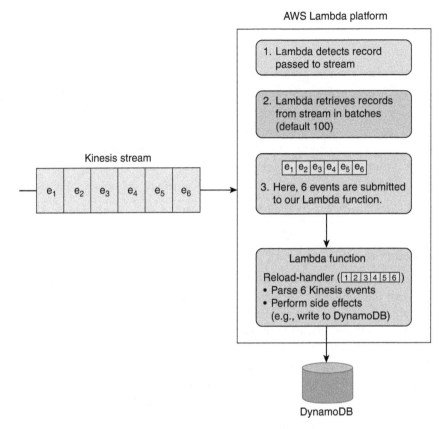

**Figure 11.7   AWS Lambda detects records posted to a Kinesis stream, collects a microbatch of those records, and then submits that microbatch to the specified Lambda function.**

and differences of Samza and Lambda are explored further in table 11.4; this table shows more similarities than we might expect.

**Table 11.4   Comparing features of Apache Samza and AWS Lambda**

| Feature | Apache Samza | AWS Lambda |
|---|---|---|
| Function invoked for | A single event | A microbatch of events |
| Executed on | Apache YARN (open source) | AWS Lambda (proprietary) |
| Events supported | Apache Kafka | Kafka and Kinesis records, S3 events, DynamoDB events, SNS notifications |
| Write function in | Java | JavaScript, Java 8, Scala (so far) |
| Store state | Locally or remotely | Remotely |

We are going to write our Lambda function in Scala, which is supported for JVM-based Lambdas alongside Java. Let's get started.

### 11.2.3 *Lambda setup and event modeling*

We need to build our Lambda function as a single fat jar file that contains all of our dependencies and can be uploaded to the Lambda service. As before, we will use Scala Build Tool, with the `sbt-assembly` plugin to build our fat jar. Let's start by creating a folder to work in, ideally inside the book's Vagrant virtual machine:

```
$ mkdir ~/aow-lambda
```

Let's add a file in the root of this folder called build.sbt, with the contents as shown in the following listing.

**Listing 11.1   build.sbt**

```
javacOptions ++= Seq("-source", "1.8", "-target", "1.8", "-Xlint")   ◁─┐  Our Lambda
                                                                         function will
lazy val root = (project in file(".")).                                  be compiled
  settings(                                                               against Scala
    name := "aow-lambda",                                                 2.12.7 on
    version := "0.1.0",                                                   Java 8.
    scalaVersion := "2.12.7",                                        ◁─┘
    retrieveManaged := true,
    libraryDependencies += "com.amazonaws" % "aws-lambda-java-core" %
  ➥ "1.2.0",
    libraryDependencies += "com.amazonaws" % "aws-lambda-java-events" %
  ➥ "2.2.4",
    libraryDependencies += "com.amazonaws" % "aws-java-sdk" % "1.11.473"
  ➥ % "provided",
    libraryDependencies += "com.amazonaws" % "aws-java-sdk-core" %
  ➥ "1.11.473" % "provided",
    libraryDependencies += "com.amazonaws" % "aws-java-sdk-kinesis" %
  ➥ "1.11.473" % "compile",
    libraryDependencies += "com.fasterxml.jackson.module" %
  ➥ "jackson-module-scala_2.12" % "2.8.4",
    libraryDependencies += "org.json4s" %% "json4s-jackson" % "3.6.2",
    libraryDependencies += "org.json4s" %% "json4s-ext" % "3.6.2",
    libraryDependencies += "com.github.seratch" %% "awscala" % "0.8.+"
  )

mergeStrategy in assembly := {                                     ◁─┐  How to handle
    case PathList("META-INF", xs @ _*) => MergeStrategy.discard        conflicts when
    case x => MergeStrategy.first                                      merging the
}                                                                     dependencies
                                                                      into our fat jar
jarName in assembly := { s"${name.value}-${version.value}" }
```

"provided" libraries are available in Lambda and so do not need to be bundled in our fat jar; "compile" ones are bundled in the fat jar.

To handle the assembly of our fat jar, we need to create a project subfolder with a plugins.sbt file within it. Create it like so:

```
$ mkdir project
$ echo 'addSbtPlugin("com.eed3si9n" % "sbt-assembly" % "0.14.9")' > \
  project/plugins.sbt
```

If you are running this from the book's Vagrant virtual machine, you should have Java 8, Scala, and SBT installed already. Check that our build is set up correctly:

```
$ sbt assembly
...
[info] Packaging /vagrant/ch11/11.2/aow-lambda/target/scala-2.12/
➥ aow-lambda-0.1.0 ...
[info] Done packaging.
[success] Total time: 860 s, completed 20-Dec-2018 18:33:57
```

Now we can move on to our event-handling code. If you look back at our analytics-on-write algorithm in section 11.1.3, you'll see that the logic is largely driven by which OOPS event type we are currently processing. So, let's first write some code to deserialize our incoming event into an appropriate Scala case class. Create the following file:

```
src/main/scala/aowlambda/events.scala
```

This file should be populated with the contents of the following listing.

---

**Listing 11.2  events.scala**

```scala
package aowlambda

import java.util.UUID, org.joda.time.DateTime
import org.json4s._, org.json4s.jackson.JsonMethods._

case class EventSniffer(event: String)
case class Employee(id: UUID, jobRole: String)
case class Vehicle(vin: String, mileage: Int)
case class Location(latitude: Double, longitude: Double, elevation: Int)
case class Package(id: UUID)
case class Customer(id: UUID, isVip: Boolean)

sealed trait Event
case class TruckArrives(timestamp: DateTime, vehicle: Vehicle,
  location: Location) extends Event
case class TruckDeparts(timestamp: DateTime, vehicle: Vehicle,
  location: Location) extends Event
case class MechanicChangesOil(timestamp: DateTime, employee: Employee,
  vehicle: Vehicle) extends Event
case class DriverDeliversPackage(timestamp: DateTime, employee: Employee,
  `package`: Package, customer: Customer, location: Location) extends Event
case class DriverMissesCustomer(timestamp: DateTime, employee: Employee,
  `package`: Package, customer: Customer, location: Location) extends Event

object Event {

  def fromBytes(byteArray: Array[Byte]): Event = {
    implicit val formats = DefaultFormats ++ ext.JodaTimeSerializers.all ++
      ext.JavaTypesSerializers.all
    val raw = parse(new String(byteArray, "UTF-8"))
    raw.extract[EventSniffer].event match {
```

Annotations:
- **Contains just the event's type, so we can determine which case class to deserialize into** (points to `case class EventSniffer(event: String)`)
- **Representing our events as an algebraic data type (ADT)** (points to `sealed trait Event`)
- **Pattern match on the event's type to construct the final Event.** (points to `raw.extract[EventSniffer].event match {`)

```
        case "TRUCK_ARRIVES" => raw.extract[TruckArrives]
        case "TRUCK_DEPARTS" => raw.extract[TruckDeparts]
        case "MECHANIC_CHANGES_OIL" => raw.extract[MechanicChangesOil]
        case "DRIVER_DELIVERS_PACKAGE" => raw.extract[DriverDeliversPackage]
        case "DRIVER_MISSES_CUSTOMER" => raw.extract[DriverMissesCustomer]
        case e => throw new RuntimeException("Didn't expect " + e)     ◁─┐
      }
    }
}
```

**In case we have an
unexpected event type**

Let's use the Scala console or REPL to check that this code is working correctly:

```
$ sbt console
scala> val bytes = """{"event":"TRUCK_ARRIVES", "location": {"elevation":7,
  "latitude":51.522834, "longitude":-0.081813},
  "timestamp": "2018-01-12T12:42:00Z", "vehicle": {"mileage":33207,
  "vin":"1HGCM82633A004352"}}""".getBytes("UTF-8")
bytes: Array[Byte] = Array(123, ...
scala> aowlambda.Event.fromBytes(bytes)
res0: aowlambda.Event = TruckArrivesEvent(2018-01-12T12:42:00.000Z,
  Vehicle(1HGCM82633A004352,33207),Location(51.522834,-0.081813,7))
```

Great—we can see that a byte array representing a *Truck arrives* event in JSON format is being correctly deserialized into our `TruckArrivesEvent` case class in Scala. Now we can move onto implementing our analytics-on-write algorithm.

### 11.2.4 *Revisiting our analytics-on-write algorithm*

Section 11.1.3 laid out the algorithm for our OOPS analytics-on-write by using simple pseudocode. This pseudocode was intended to process a single event at a time, but we have since learned that our Lambda function will be given a microbatch of up to 100 events at a time. Of course, we could apply our algorithm to each event in the microbatch, as shown in figure 11.8.

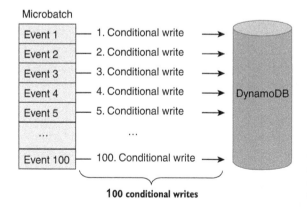

Figure 11.8  A naïve approach to updating our DynamoDB table from our microbatch of events would involve a conditional write for every single event.

Unfortunately, this approach is wasteful whenever a single microbatch contains multiple events relating to the same OOPS truck, as could easily happen. Remember how we want to update our DynamoDB table with a given truck's most recent (highest) mileage? If our microbatch contains 10 events relating to Truck 123, the naïve implementation of our algorithm from figure 11.8 would require 10 conditional writes to the DynamoDB table just to update the Truck 123 row. Reading and writing to remote databases is an expensive operation, so it's something that we should aim to minimize in our Lambda function, especially given that each Lambda function call has to complete within 60 seconds (configurable up to 15 minutes).

The solution is to *pre-aggregate* our microbatch of events inside our Lambda function. With 10 events in the microbatch relating to Truck 123, our first step should be to find the highest mileage across these 10 events. This highest mileage figure is the only one that we need to attempt to write to DynamoDB; there is no point bothering DynamoDB with the nine lower mileages at all. Figure 11.9 depicts this pre-aggregation technique.

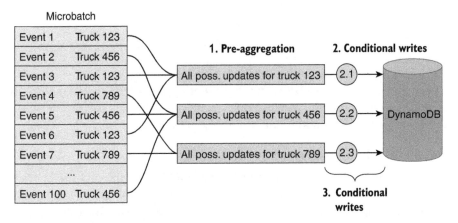

**Figure 11.9   By applying a pre-aggregation to our microbatch of events, we can reduce the required number of conditional writes down to one per OOPS truck found in our microbatch.**

Of course, the pre-aggregation step may not always bear fruit. For example, if all 100 events in the microbatch relate to different OOPS trucks, there is nothing to reduce. This said, the pre-aggregation is a relatively cheap in-memory process, so it's worth always attempting it, even if we prevent only a handful of unnecessary DynamoDB operations.

What format should our pre-aggregation step output? Exactly the same as the format of our data in DynamoDB! The pre-aggregation step and the in-Dynamo aggregation differ from each other in terms of their scope (100 events versus all-history) and their storage mechanism (local in-memory versus remote database), but they represent the same analytics algorithm, and a different intermediate format would only confuse our OOPS colleagues. Table 11.5 reminds us of the format we need to populate in our Lambda function.

**Table 11.5 Our DynamoDB row layout dictates the format of our pre-aggregated row in our Lambda function.**

| Truck VIN | Latitude | Longitude | Location timestamp | Mileage | Mileage at oil change |
|-----------|----------|-----------|--------------------|---------|-----------------------|
| 1HGCM8... | 51.5... | -0.15... | 2018-08-02T21:50:49Z | 12453 | 12418 |

We should be ready now to create our first analytics-on-write code for this AWS Lambda. Create the following file:

```
src/main/scala/aowlambda/aggregator.scala
```

This file should be populated with the contents of the following listing.

**Listing 11.3 aggregator.scala**

```scala
package aowlambda

import org.joda.time.DateTime, aowlambda.{TruckArrives => TA},
  aowlambda.{TruckDeparts => TD}, aowlambda.{MechanicChangesOil => MCO}

case class Row(vin: String, mileage: Int, mileageAtOilChange: Option[Int],
  locationTs: Option[(Location, DateTime)])          ⟵ The intermediate format for a truck's metrics

object Aggregator {

  def map(event: Event): Option[Row] = event match {          ⟵ Transforms an event into the intermediate format
    case TA(ts, v, loc) => Some(Row(v.vin, v.mileage, None, Some(loc, ts)))
    case TD(ts, v, loc) => Some(Row(v.vin, v.mileage, None, Some(loc, ts)))
    case MCO(ts, _, v)  => Some(Row(v.vin, v.mileage, Some(v.mileage), None))
    case _              => None
  }

  def reduce(events: List[Option[Row]]): List[Row] =          ⟵ Reduces all events into one row per truck
    events
      .collect { case Some(r) => r }
      .groupBy(_.vin)
      .values
      .toList
      .map(_.reduceLeft(merge))

  private val merge: (Row, Row) => Row = (a, b) => {          ⟵ Consolidates two rows into one by taking the newer/higher values

    val m = math.max(a.mileage, b.mileage)
    val maoc = (a.mileageAtOilChange, b.mileageAtOilChange) match {
      case (l @ Some(_), None) => l
      case (l @ Some(lMaoc), Some(rMaoc)) if lMaoc > rMaoc => l
      case (_, r) => r
    }
    val locTs = (a.locationTs, b.locationTs) match {
      case (l @ Some(_), None) => l
      case (l @ Some((_, lTs)), Some((_, rTs))) if lTs.isAfter(rTs) => l
```

The intermediate format for a truck's metrics

```
    . case (_, r) => r
    }
    Row(a.vin, m, maoc, locTs)
  }
}
```

There's quite a lot to unpack in our aggregator.scala file; let's make sure you understand what it's doing before moving on. First, we have the imports, including aliases for our event types to make the subsequent code fit on one line. Then we introduce a case class called Row; this is a lightly adapted version of the data held in table 11.5. Figure 11.10 shows the relationship between table 11.5 and our new Row case class; we will be using instances of this Row to drive our updates to DynamoDB.

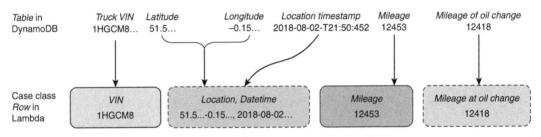

**Figure 11.10   The Row case class in our Lambda function will contain the same data points as found in our DynamoDB table. The dotted lines indicate that a given Row instance may not contain these data points, because the related event types were not found in this microbatch for this OOPS truck.**

Moving onto our Aggregator object, the first function we see is called map. This function transforms any incoming OOPS event into either Some(Row) or None, depending on the event type. We use a Scala pattern match on the event type to determine how to transform the event:

- The three event types with relevant data for our analytics are used to populate different slots in a new Row instance.
- Any other event types simply output a None, which will be filtered out later.

If you are unfamiliar with the Option boxing of the Row, you can find more on these techniques in chapter 8.

The second public function in the Aggregator is called reduce. This takes a list of Option-boxed Rows and squashes them down into a hopefully smaller list of Rows. It does this by doing the following:

- Filtering out any Nones from the list
- Grouping the Rows by the VIN of the truck described in that Row
- Squashing down each group of Rows for a given truck into a single Row representing every potential update from the microbatch for this truck

Phew! If this seems complicated, it's because it *is* complicated. We are rolling our own map-reduce algorithm to run against each microbatch of events in this Lambda. With the sophisticated query languages offered by technologies such as Apache Hive, Amazon Redshift, and Apache Spark, it's easy to forget that, until relatively recently, coding MapReduce algorithms for Hadoop in Java was state-of-the-art. You could say that with our Lambda, we are getting back to basics in writing our own bespoke map-reduce code.

How do we squash a group of Rows for a given truck down to a single Row? This is handled by our reduceLeft, which takes pairs of Rows and applies our merge function to each pair iteratively until only one Row is left. The merge function takes two Rows and outputs a single Row, combining the most recent data-points from each source Row. Remember, the whole point of this pre-aggregation is to minimize the number of writes to DynamoDB. Figure 11.11 shows the end-to-end map-reduce flow.

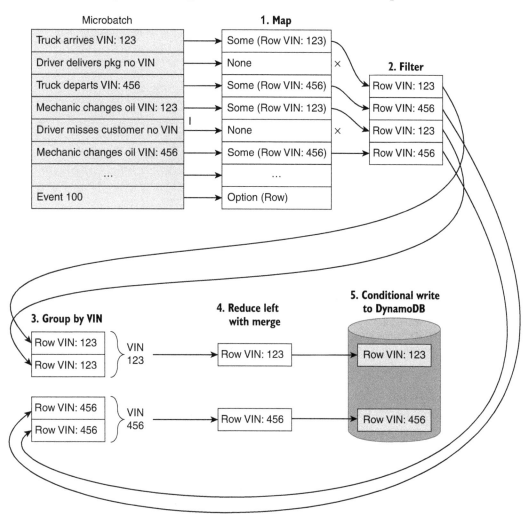

**Figure 11.11**  First, we map the events from our microbatch into possible Rows, and filter out the Nones. We then group our Rows by the truck's VIN and reduce each group to a single Row per truck by using a merge. Finally, we perform a conditional write for each Row against our table in DynamoDB.

That completes our map-reduce code for pre-aggregating our microbatch of events prior to updating DynamoDB. In the next section, let's finally write that DynamoDB update code.

### 11.2.5  Conditional writes to DynamoDB

Thanks to our merge function, we know that the reduced list of Rows represents the most recent data-points for each truck *from within the current microbatch*. But it is possible that another microbatch with more recent events for one of these OOPS trucks has already been processed. Event transmission and collection is notoriously unreliable, and it's possible that our current microbatch contains old events that got delayed somehow—perhaps because an OOPS truck was in a road tunnel, or because the network was down in an OOPS garage. Figure 11.12 illustrates this risk of our microbatches being processed out of order.

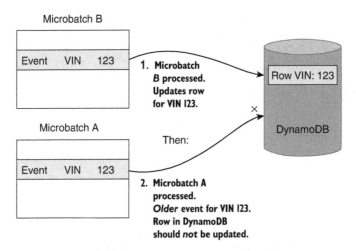

**Figure 11.12   When our Lambda function processes events for OOPS Truck 123 out of chronological order, it is important that a more recent data point in DynamoDB is not overwritten with a stale data point.**

As a result, we cannot blindly overwrite the existing row in DynamoDB with the new row generated by our Lambda function. We could easily overwrite a more recent truck mileage or location with an older one! The solution has been mentioned briefly in previous sections of this chapter; we are going to use a feature of DynamoDB called *conditional writes*.

When I alluded to conditional writes earlier in this chapter, I suggested that there was some kind of read-check-write loop performed by the Lambda against DynamoDB. In fact, things are simpler than this: all we have to do in our Lambda is send each write request to DynamoDB with a condition attached to it; DynamoDB will then check the condition against the current state of the database, and apply the write only if the condition passes. Figure 11.13 shows the set of conditional writes our Lambda will attempt to perform for each row.

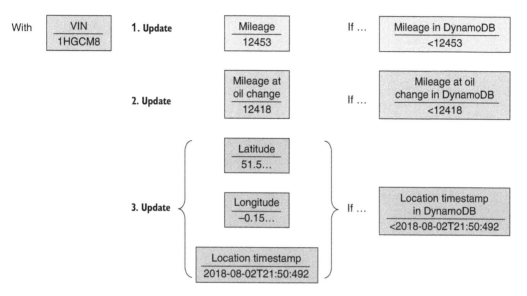

**Figure 11.13** For each Row, we attempt three writes against DynamoDB, where each write is dependent on a condition in DynamoDB passing.

We are ready to take the logic from figure 11.13 and implement it in Scala. For this, we will use the AWS Java SDK, which includes a DynamoDB client. To keep our code succinct, we will also use the AWScala project, which is a more Scala-idiomatic domain-specific language (DSL) for working with DynamoDB.

Create this Scala file:

```
src/main/scala/aowlambda/Writer.scala
```

This file should be populated with the contents of the following listing.

**Listing 11.4  Writer.scala**

```
package aowlambda

import awscala._, dynamodbv2.{AttributeValue => AttrVal, _}
import com.amazonaws.services.dynamodbv2.model._
import scala.collection.JavaConverters._

object Writer {
  private val ddb = DynamoDB.at(Region.US_EAST_1)

  private def updateIf(key: AttributeValue, updExpr: String,
    condExpr: String, values: Map[String, AttributeValue],
    names: Map[String, String]) {
```

Perform a conditional write using the supplied arguments.

```scala
    val updateRequest = new UpdateItemRequest()
      .withTableName("oops-trucks")
      .addKeyEntry("vin", key)
      .withUpdateExpression(updExpr)
      .withConditionExpression(condExpr)
      .withExpressionAttributeValues(values.asJava)
      .withExpressionAttributeNames(names.asJava)

    try {
      ddb.updateItem(updateRequest)
    } catch { case ccfe: ConditionalCheckFailedException => }
  }

  def conditionalWrite(row: Row) {
    val vin = AttrVal.toJavaValue(row.vin)

    updateIf(vin, "SET #m = :m",
      "attribute_not_exists(#m) OR #m < :m",
      Map(":m" -> AttrVal.toJavaValue(row.mileage)),
      Map("#m" -> "mileage"))

    for (maoc <- row.mileageAtOilChange) {
      updateIf(vin, "SET #maoc = :maoc",
        "attribute_not_exists(#maoc) OR #maoc < :maoc",
        Map(":maoc" -> AttrVal.toJavaValue(maoc)),
        Map("#maoc" -> "mileage-at-oil-change"))
    }

    for ((loc, ts) <- row.locationTs) {
      updateIf(vin, "SET #ts = :ts, #lat = :lat, #long = :long",
        "attribute_not_exists(#ts) OR #ts < :ts",
        Map(":ts"   -> AttrVal.toJavaValue(ts.toString),
            ":lat"  -> AttrVal.toJavaValue(loc.latitude),
            ":long" -> AttrVal.toJavaValue(loc.longitude)),
        Map("#ts"   -> "location-timestamp", "#lat" -> "latitude",
            "#long" -> "longitude"))
    }
  }
}
```

**If the condition check fails, an Exception is thrown, which we will silently swallow.**

**Update the truck's mileage if the truck's current mileage is missing or lower than the new value.**

**Update the three location fields if they are supplied and the timestamp in DynamoDB is lower.**

**Update the miles at oil change if it is supplied and the figure in DynamoDB is lower.**

The code for interfacing with DynamoDB is admittedly verbose, but squint at this code a little and you can see that it follows the basic shape of figure 11.13. We have a Writer module that exposes a single method, conditionalWrite, which takes in a Row as its argument. The method does not return anything; it exists only for the side effects it performs, which are conditional writes of the three elements of the Row against the same truck's row in DynamoDB. These conditional writes are performed using the private updateIf function, which helps construct an UpdateItemRequest for the row in DynamoDB.

A few things to note about this code:

- The updateIf function takes the truck's VIN, the update statement, the update condition (which must pass for the update to be performed), the values required for the update, and attribute aliases for the update.

- The update statement and update condition are written in DynamoDB's custom expression language,[1] which uses `:some-value` for attribute values and `#some-alias` for attribute aliases.
- For mileage at oil change and the truck location, those conditional writes are themselves dependent on the corresponding parts of the `Row` being populated.
- For brevity, we do not have any logging or error handling. In a production Lambda function, we would include both.

That completes our DynamoDB-specific code!

### 11.2.6  *Finalizing our Lambda*

Now we are ready to pull our logic together into the Lambda function definition itself. Create our fourth and final Scala file:

```
src/main/scala/aowlambda/LambdaFunction.scala
```

This file should be populated with the contents of the following listing.

##### Listing 11.5  LambdaFunction.scala

```scala
package aowlambda

import com.amazonaws.services.lambda.runtime.events.KinesisEvent
import scala.collection.JavaConverters._

class LambdaFunction {

  def recordHandler(microBatch: KinesisEvent) {          // Converts our microbatch of
                                                         // events into a List of Rows
    val allRows = for {                                  // ready for reduction
      recs <- microBatch.getRecords.asScala.toList
      bytes = recs.getKinesis.getData.array
      event = Event.fromBytes(bytes)
      row = Aggregator.map(event)
    } yield row
                                                         // Reduces our Rows to the
    val reducedRows = Aggregator.reduce(allRows)         // minimal set of possible
                                                         // updates for DynamoDB

    for (row <- reducedRows) {                           // Loops through each Row
      Writer.conditionalWrite(row)                       // and performs a conditional
    }                                                    // write in DynamoDB
  }
}
```

---

[1]  For more information on Dynamo DB's expression language, refer to https://docs.aws.amazon.com/amazon-dynamodb/latest/developerguide/Expressions.html.

It is this function, `aowlambda.LambdaFunction.recordHandler`, that will be invoked by AWS Lambda for each incoming microbatch of Kinesis events. The code should be fairly simple:

1  We convert each Kinesis record in the microbatch into a `Row` instance.
2  We run our reduce function to pre-aggregate all of our `Row`s down to the minimal set of DynamoDB updates to attempt (see listing 11.3).
3  We loop through the remaining `Row`s and perform a DynamoDB conditional write for each (see listing 11.4).

Code complete! Let's quickly check that the code compiles fine, like so:

```
guest$ sbt compile
...
[info] Compiling 2 Scala sources to /vagrant/ch11/11.2/aow-
    lambda/target/scala-2.11/classes...
[success] Total time: 19 s, completed 20-Dec-2018 21:00:45
```

Now we need to assemble a fat jar with all of our Lambda function's dependencies in it. Let's use the `sbt-assembly` plugin for Scala Build Tool to do this:

```
guest$ sbt assembly
...
[info] Packaging /vagrant/ch11/11.2/aow-lambda/target/scala-2.11/
➥ aow-lambda-0.1.0 ...
[info] Done packaging.
[success] Total time: 516 s, completed 20-Dec-2018 21:10:04
```

Great—our build has now been completed. This completes the coding required for our Lambda function. In the next section, we will get this deployed and put it through its paces!

## 11.3  Running our Lambda function

Much of the appeal of a system like AWS Lambda lies in its "serverless" promise. Of course, servers *are* involved in running our Lambda, but they are hidden from us, with AWS taking responsibility for operating our function reliably. The price we pay for outsourcing our ops is a relatively involved setup process, which we will go through now.

### 11.3.1  Deploying our Lambda function

Earlier in this chapter, we set up our Kinesis stream and our DynamoDB table. We now need to deploy our assembled Lambda function and wire it into our stream and table. We will perform all of this by using the AWS CLI tools. There is a lot of ceremony to step through here, so let's get started.

### UPLOADING TO S3

First, we need to make our local fat jar accessible to the AWS Lambda service. We do this by uploading the file to Amazon S3. This is straightforward using the AWS CLI. First, we create the bucket:

```
$ s3_bucket=ulp-ch11-fatjar-${your_first_pets_name}
$ jar=aow-lambda-0.1.0
$ aws s3 mb s3://${s3_bucket} --profile=ulp --region=us-east-1
make_bucket: s3://ulp-ch11-fatjar-little-torty/
```

And next we upload the jar file to S3:

```
$ aws s3 cp ./target/scala-2.12/${jar} s3://${s3_bucket}/ --profile=ulp
upload: target/scala-2.12/aow-lambda-0.1.0 to
 s3://ulp-ch11-fatjar-little-torty/aow-lambda-0.1.0
```

Unfortunately, there is a bug with AWS whereby uploading large files to newly created buckets can hang. If that happens, take a break and try again in an hour or so; you can continue with the rest of the setup in the meantime. Next, we need to configure the necessary permissions for our Lambda function.

### CONFIGURING PERMISSIONS

The permissions required for operating a Lambda function are complex, so we are going to take a shortcut by using a CloudFormation template that we prepared earlier. AWS CloudFormation is a service that lets you spin up various AWS resources by using JSON templates; you can think of a CloudFormation template as the declarative JSON recipe for creating a collection of AWS services configured exactly as you want them. In Amazon parlance, we will use this template to create a *stack* of AWS resources—in our case, IAM roles to operate our Lambda function.

The pre-prepared CloudFormation template is publicly available in S3 in the ulp-assets bucket:

```
$ template=https://ulp-assets.s3.amazonaws.com/ch11/cf/aow-lambda.template
```

Kick off the stack creation like so:

```
$ aws cloudformation create-stack --stack-name AowLambda \
  --template-url ${template} --capabilities CAPABILITY_IAM \
  --profile=ulp --region=us-east-1
{
    "StackId": "arn:aws:cloudformation:us-east-1:719197435995:
  stack/AowLambda/392e05e0-5963-11e5-aa74-5001ba48c2d2"
}
```

You can monitor Amazon's progress in creating this stack by using this command:

```
$ aws cloudformation describe-stacks --stack-name AowLambda \
  --profile=ulp --region=us-east-1
```

When you see a StackStatus of CREATE_COMPLETE in the returned JSON, we are ready to continue. We will need another piece of this output in the next section: the Output-Value for the ExecutionRole, which should start with arn:aws:iam::. Let's create a new environment variable set to this value:

```
$ role_arn="arn:aws:iam::719197435995:role/AowLambda-LambdaExecRole
➥ -1CNLT4WVY6PN4"
```

When creating this variable, make sure not to include any line breaks in the value.

### CREATING OUR LAMBDA FUNCTION

The next step is to register our function with AWS Lambda. We can do this with a single AWS CLI command:

```
$ aws lambda create-function --function-name AowLambda \
  --role ${role_arn} --code S3Bucket=${s3_bucket},S3Key=${jar} \
  --handler aowlambda.LambdaFunction::recordHandler \
  --runtime java8 --timeout 60 --memory-size 1024 \
  --profile=ulp --region=us-east-1
  {
    "FunctionName": "AowLambda",
    "FunctionArn": "arn:aws:lambda:us-east-1:089010284850:function:
➥ AowLambda",
    "Runtime": "java8",
    "Role": "arn:aws:iam::089010284850:role/AowLambda-LambdaExecRole
➥ -FUVSBSEC1Y6R",
    "Handler": "aowlambda.LambdaFunction::recordHandler",
    "CodeSize": 27765196,
    "Description": "",
    "Timeout": 60,
    "MemorySize": 1024,
    "LastModified": "2018-12-21T07:44:31.073+0000",
    "CodeSha256": "jRpr4E6OrP4hznB1Q/ApO6+fOAnLHMwfyhhT3rU5KWM=",
    "Version": "$LATEST",
    "TracingConfig": {
      "Mode": "PassThrough"
    },
    "RevisionId": "0609f559-fd1f-45c9-aee2-11ee159183b5"
  }
```

This command defines a function called AowLambda by using our IAM role and our fat jar previously uploaded to S3. The --handler argument tells AWS Lambda exactly which method to invoke inside our fat jar. The next three arguments configure the exact operation of the function: the function should be run against Java 8 (as opposed to Node.js), should time out after 60 seconds, and should be given 1 GB of RAM.

### ATTACHING OUR FUNCTION TO KINESIS

Are we there yet? Not quite. We still need to identify the Kinesis stream we created earlier as the event source for our Lambda function. We do this by creating an *event source mapping* between the Kinesis stream and the function, again using the AWS CLI:

```
$ aws lambda create-event-source-mapping \
  --event-source-arn ${stream_arn} \
  --function-name AowLambda --enabled --batch-size 100 \
  --starting-position TRIM_HORIZON --profile=ulp --region=us-east-1
{
    "UUID": "bdf15c0b-a565-4a15-b790-6c2247d9aba3",
    "StateTransitionReason": "User action",
    "LastModified": 1545378677.527,
    "BatchSize": 100,
    "EventSourceArn": "arn:aws:kinesis:us-east-1:719197435995:stream/
⟹ oops-events",
    "FunctionArn": "arn:aws:lambda:us-east-
    1:719197435995:function:AowLambda",
    "State": "Creating",
    "LastProcessingResult": "No records processed"
}
```

From the returned JSON, you can see that we have successfully connected our Lambda function to our stream of OOPS events. It reports that no records have been processed yet. In the next section, we will re-enable our event generator and check the results.

### 11.3.2 Testing our Lambda function

Remember that our generator script is available in the GitHub repository:

```
ch11/11.1/generate.py
```

Let's kick off our event generator:

```
$ /vagrant/ch11/11.1/generate.py
Wrote DriverDeliversPackage with timestamp 2018-01-01 02:31:00
Wrote DriverMissesCustomer with timestamp 2018-01-01 05:53:00
Wrote TruckDepartsEvent with timestamp 2018-01-01 08:21:00
```

Great! Those events are now being sent to our Kinesis stream. Let's now take a look at our DynamoDB table. From the AWS dashboard:

- Make sure you are in the N. Virginia region by using the top-right drop-down menu.
- Click DynamoDB.
- Click the `oops-truck` table entry and click the Explore Table button.

You should see three rows with all of the expected fields, as shown in figure 11.14.

Leave our event generator running a little longer and then refresh the table in the DynamoDB interface. You should see the following:

- We now have values populated for all attributes for all three trucks. In fact, our generator sends only events for three OOPS trucks.
- Our location timestamps roughly match the most recent timestamps we see printed in the terminal running our generator, showing that our Lambda function is keeping up.

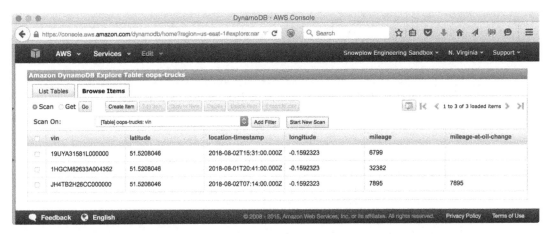

**Figure 11.14   Our DynamoDB table oops-trucks contains the six fields expected by the OOPS business intelligence team. Note that the Lambda has observed an oil change event for only one of the trucks so far.**

- The trucks' latitude and longitude have changed. Expect to see the same latitudes and longitudes repeated frequently, as the generator uses only five locations.
- The mileage-at-oil-change for each truck lags the truck's total mileage.

Figure 11.15 shows this updated view we are expecting to see in DynamoDB.

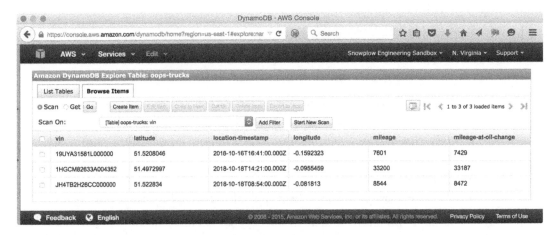

**Figure 11.15   We now have all values populated for each of our three trucks. These values are being constantly updated by our AWS Lambda function in response to new events.**

So far, so good—we can see that our Lambda is working well to keep our operational dashboard in DynamoDB up-to-date with the latest stream of events from OOPS trucks.

The last thing we want to check is that our conditional writes are correctly guarding against old events that could arrive out of order. Remember that we don't want to overwrite a fresh data point in DynamoDB with older data just because the older event arrives after the newer event.

To test this, first refresh the DynamoDB table to get the latest data. Then switch back to your generator's terminal window and press Ctrl-C:

```
Wrote TruckArrivesEvent with timestamp 2018-03-06 00:39:00
^CTraceback (most recent call last):
  File "./generate.py", line 168, in <module>
    time.sleep(1)
KeyboardInterrupt
```

Finally, we restart the generator with a new command-line argument, backward:

```
$ ./generate.py backwards
Wrote DriverDeliversPackage with timestamp 2017-12-31 20:15:00
Wrote TruckArrivesEvent with timestamp 2017-12-31 18:40:00
Wrote MechanicChangesOil with timestamp 2017-12-31 15:45:00
```

As you can see, the generator is now working backward, creating ever-older events. Leave it a short while, and then head back into the DynamoDB interface and refresh the table. You'll see that all of the data points are unchanged: the old events are not impacting on our latest truck statuses in DynamoDB at all.

For a final test, leave the backward-stepping event generator running, open a new terminal, and kick off the forward-stepping generator again:

```
$ ./generate.py
Wrote DriverDeliversPackage with timestamp 2018-01-01 00:38:00
Wrote TruckArrivesEvent with timestamp 2018-01-01 05:03:00
Wrote MechanicChangesOil with timestamp 2018-01-01 08:41:00
```

You will have to wait a while until the event timestamps that are generated overtake your current location-timestamp values in DynamoDB, but after this happens, you should see your truck data points in DynamoDB starting to refresh again.

And that completes our testing! Hopefully, our colleagues at OOPS are pleased with the result: an analytics-on-write system that can keep a dashboard of OOPS trucks updated in near real-time from a Kinesis stream. A nice evolution for this project would be to add a visualization layer on top of the DynamoDB table, perhaps implemented in D3.js or a similar library.

## *Summary*

- With analytics-on-write, we decide on the analysis we want to perform ahead of time, and then put this analysis live on our event stream.
- Analytics-on-write is good for low-latency operational reporting and dashboards to support thousands of simultaneous users. Analytics-on-write is a complement to analytics-on-read, not a competitor.
- To support the latency and access requirements, analytics-on-write systems typically lean heavily on horizontally scalable key-value stores such as Amazon DynamoDB.

- Applying analytics-on-write retrospectively can be difficult, so it is important to gather requirements and agree on the proposed algorithm up-front.
- We implemented a simple truck status dashboard for OOPS by using AWS Lambda as the stream processing framework reading OOPS events from Amazon Kinesis.
- To reduce the number of calls to DynamoDB, we implemented a local map-reduce on the microbatch of events received by the Lambda function, reducing this batch of 100 events down to the minimal number of potential updates to DynamoDB.
- AWS Lambda is essentially stateless, so we used DynamoDB's conditional writes to ensure that we were always updating our DynamoDB table with the latest data points relating to each OOPS truck.
- There was a close mapping between the dashboard required by OOPS, the layout of the table in DynamoDB, and the Scala representation of a row used in our Lambda's local map-reduce.

# *appendix*
# *AWS primer*

This appendix provides a brief primer on Amazon Web Services to get you up to speed with the AWS environment and services.

Please note that, because Amazon Kinesis Data Streams is not currently available in AWS Free Tier, the procedures in this book necessarily involve creating live resources in your Amazon Web Services account, which can incur some charges.[1] Don't worry—we will tell you as soon as you can safely delete a given resource. In addition, you can also set alerts on your spending in order to be notified whenever the charges go above a certain threshold.[2]

Amazon Kinesis is a fully hosted service, available only to users of the Amazon Web Services platform. Don't worry if you haven't worked with AWS before. This appendix introduces the key building blocks of AWS and will help you get set up on the platform.

---

[1] For detailed pricing information on the AWS Kinesis Data Streams service, see https://aws.amazon.com/kinesis/streams/pricing/.

[2] You can read how to set spending alerts on your AWS usage here: https://docs.aws.amazon.com/awsaccountbilling/latest/aboutv2/billing-getting-started.html#d0e1069.

## A.1 Setting up the AWS account

To get the most out of this book, you need an AWS account. If you don't have one, you can sign up by clicking the Get Started for Free or Create a Free Account button on the AWS homepage:

```
https://aws.amazon.com/
```

If you are planning on working through these examples in your company's own AWS account, we highly recommend instead that you ask your company to create a new AWS account, called Developers' Sandbox or something similar, and connect it to your main company AWS account by using Consolidated Billing.[3] This way, you can experiment in the sandbox safe in the knowledge that you cannot impact (for example, accidentally delete) existing resources in any way.

After you are signed up, log in to AWS and you should see a dashboard something like the one in figure A.1. The AWS offering is something of a zoo; by the time you read this, the dashboard may even display some new services. We've highlighted the services we will be using in this section; see table A.1 for a brief rundown on each of the highlighted services.

**Table A.1  AWS Services we'll be using in this book**

| Service | Short form | Description |
|---------|-----------|-------------|
| Identity & Access Management | `iam` | For securely controlling access to AWS services and resources for your users |
| Kinesis | `kinesis` | A fully managed unified log service |

Before we dive into Kinesis, we will first configure some sensible security settings by using Amazon's Identity & Access Management service. Let's get started.

---

[3] You can read more about how to set up consolidated billing here: https://docs.aws.amazon.com/awsaccount-billing/latest/aboutv2/consolidated-billing.html.

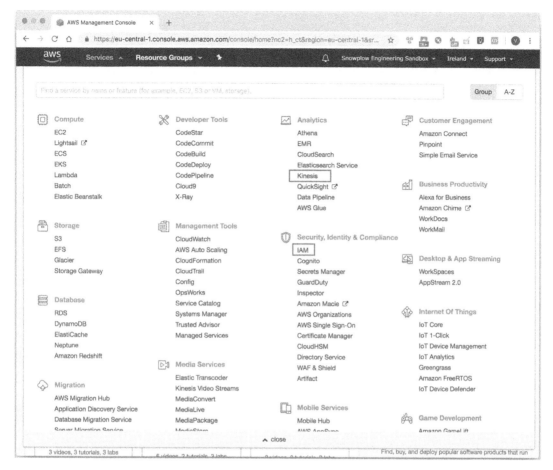

**Figure A.1** In this book, we will be working with several AWS services, including Kinesis and Identity & Access Management. These are shown highlighted in the AWS dashboard. We will also use other services such as Redshift, S3, and Elastic MapReduce, but these will be introduced directly in the corresponding chapters.

## A.2 Creating a user

As a first step, we are going to use Identity & Access Management (IAM) to create a user with the permissions on AWS resources that we will need for this book. From the AWS dashboard, follow these steps:

1 Click the Identity & Access Management icon.
2 Click Users in the left-hand navigation pane.
3 Click the Add User button.

The next screen is a four-step wizard for creating a user. Go ahead and add in a new user, called `ulp` for *unified log processing*, and make sure to select the Programmatic Access option, as shown in figure A.2. Then click the Next: Permissions button.

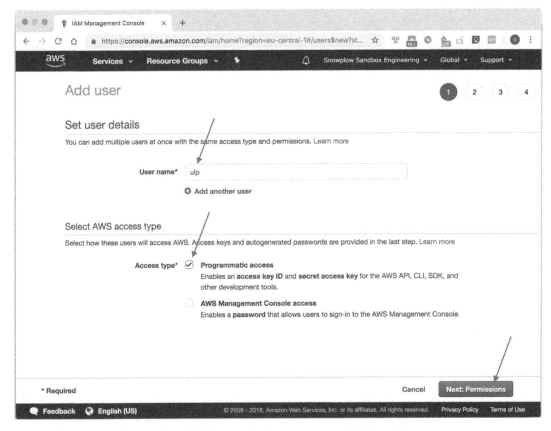

**Figure A.2    We create a new IAM user called `ulp` in the AWS user interface, making sure to select the Programmatic Access option.**

On the next screen, shown in figure A.3, you need to set the permissions for the `ulp` user. You need to attach what AWS calls a *managed policy* so that our new `ulp` user can create new streams in Kinesis, as well as write to and read from those streams. To do this:

1  Click the Attach Existing Policies Directly button.
2  In the search box, type `AmazonKinesisFullAccess`.
3  Select the policy called AmazonKinesisFullAccess.
4  Click the Next: Review button.

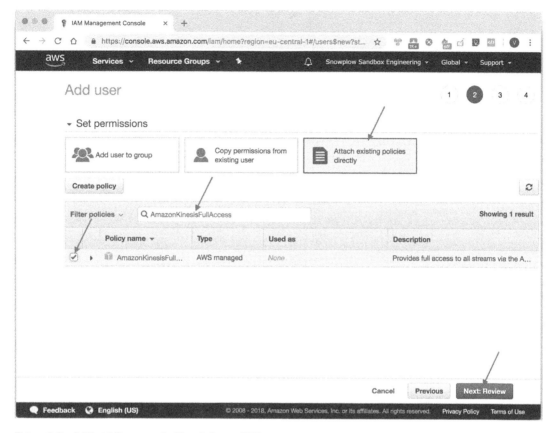

**Figure A.3  Adding full access to Kinesis to our IAM user**

These permissions on Kinesis are more generous than they need to be, but they will help you get started with Kinesis with a minimum of fuss. You should reduce the scope of these permissions later as you get more comfortable with AWS.

On the next screen, you can review the details before effectively creating the user, as shown in figure A.4.

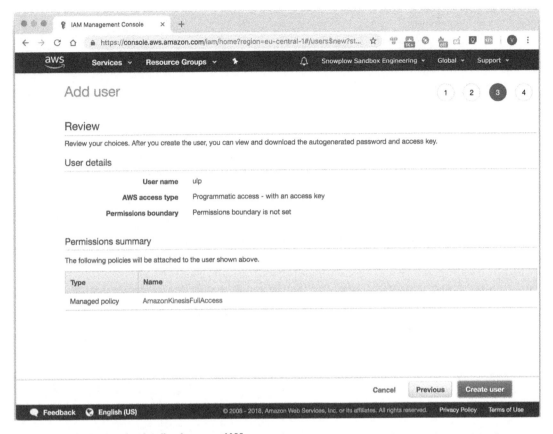

**Figure A.4   Reviewing the details of our new IAM user**

If everything seems correct, click the Create User button. This brings us to the next screen, shown in figure A.5. This screen shows a set of *user security credentials*—consisting of an Access Key ID and a Secret Access key. You can think of these as the username and password for access to the various AWS APIs.

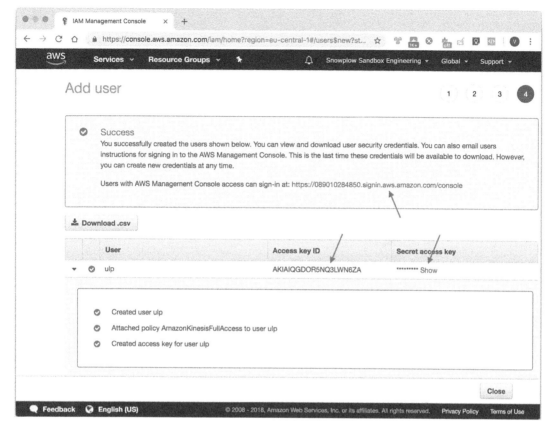

**Figure A.5** The user security credentials for our new IAM user consist of an Access Key ID and a Secret Access Key.

Make sure to download these or otherwise note them, as you will need these shortly to set up the AWS CLI. Note that the Secret Access Key is hidden, and make sure to reveal it fully by clicking Show. Also note the link with which you'll be able to access the AWS Management Console by using the `ulp` user later.

Now click Close, click Users in the left navigation pane again, and then click your new `ulp` user. The screen should look like figure A.6.

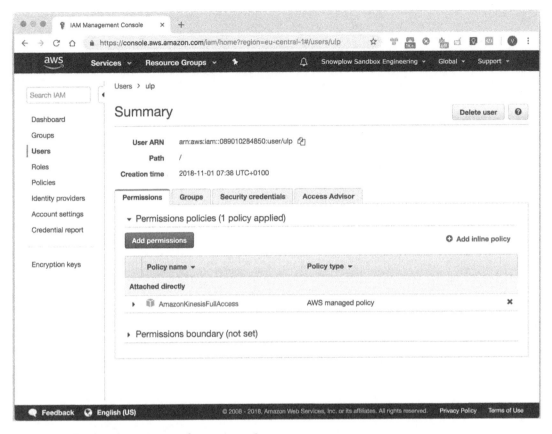

**Figure A.6   The management screen for your new `ulp` user**

Next, you need to give our `ulp` user a password so that you can access the AWS dashboard as this user, rather than as our all-powerful (and thus dangerous) root administrator. Click the Security Credentials tab, and then, on the line that reads Console Password, click the Manage link. A pop-up appears, as shown in figure A.7. Enable Console Access, leave the Autogenerated Password option selected, and click Apply.

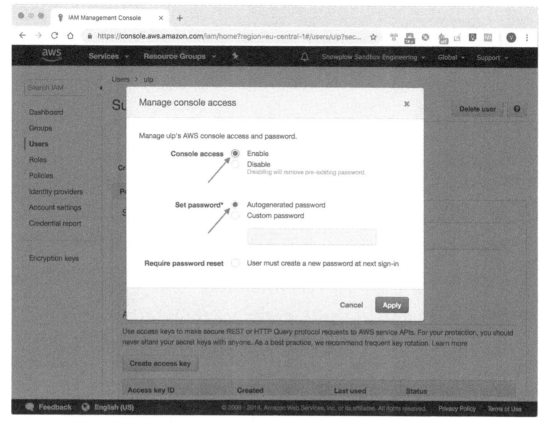

**Figure A.7  Setting up an autogenerated password for your new `ulp` user**

A new pop-up appears, showing the generated password. Click the Show option, and make sure to note the password shown under User Security Credentials, as illustrated in figure A.8. Then close the pop-up.

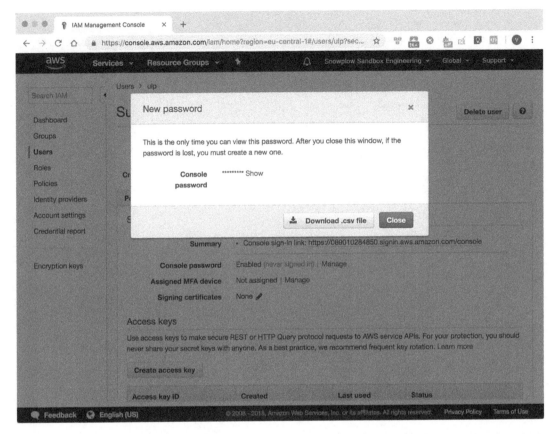

**Figure A.8    The password for our `ulp` user**

As a final step, let's log out of AWS by using the Sign Out option in the top-right drop-down, and then log back in as our new `ulp` user by using the link shown in figure A.4 (also present in the Security Credentials tab), not forgetting the password you wrote down earlier. Now click from the dashboard through to the Kinesis service, and you should see a screen like figure A.9. The prominent Create Kinesis Stream button tells you that you should have all the appropriate permissions to work with Kinesis.

With our user set up with appropriate security credentials and Kinesis permissions, the next step is to configure the AWS CLI.

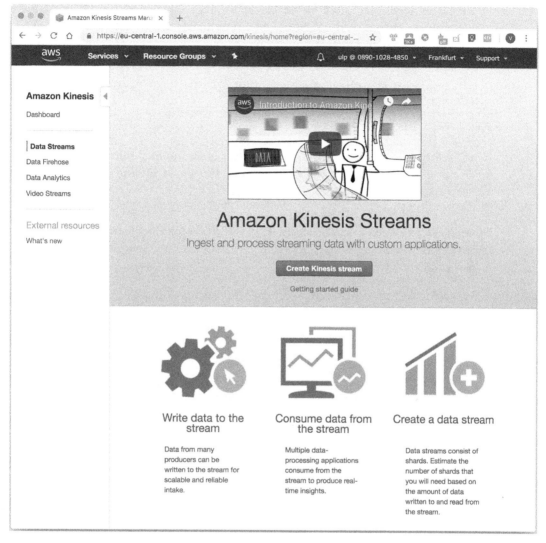

**Figure A.9** When you do not yet have any Kinesis streams set up, clicking the Kinesis icon on the AWS dashboard takes you to this Amazon Kinesis Streams screen, with a prominent Create Kinesis Stream button.

## A.3 Setting up the AWS CLI

Although the AWS web interface is easy to use, where possible in this book we will work with AWS resources using the official AWS command-line interface (CLI) application. The AWS CLI isn't much to look at, but its command syntax is quite intuitive, and it tends to get new AWS features before the web interface. The team behind the AWS CLI is also relatively good at maintaining backward compatibility, so hopefully these instructions will continue to work for the foreseeable future!

First, you need to get hold of the CLI application. If you are running this book's development environment in Vagrant, you are in luck—the AWS CLI is preinstalled. Check that it's available and working by navigating to your local copy of the Unified-Log-Processing repository and typing this:

```
host$ vagrant up && vagrant ssh
guest$ aws
usage: aws [options] <command><subcommand> [parameters]
aws: error: too few arguments
```

If you are not using the prepackaged Vagrant environment, this page in the user guide has all the information you need to install the AWS CLI onto your system:

```
https://docs.aws.amazon.com/cli/latest/userguide/installing.html
```

Done? Next, you need to configure an AWS CLI *profile* for all of your work. By default, the AWS CLI will use an implicit global profile, but it's safer to work with an explicit, named profile. So type in the following:

```
$ aws configure --profile=ulp
```

When prompted, fill in the requested details, supplying the Access Key ID and Secret Access Key you saved earlier:

```
AWS Access Key ID [None]: AKIAIWSMFSNA2ZH6W4UQ
AWS Secret Access Key [None]: uOGIOXssDw/ZtzXxxXxXXxpQvgB3Dus0zFnywWr9
Default region name [eu-west-1]: us-east-1
Default output format [None]:
```

For this appendix, we are going to assume that you chose us-east-1 as your default AWS region, but it's fine to choose another region if you prefer; just remember to update the code accordingly. You are now ready to start experimenting with Amazon Kinesis by using the AWS CLI.

# index

# RELATED MANNING TITLES

### Reactive Design Patterns

by Roland Kuhn
 with Brian Hanafee and Jamie Allen

ISBN: 9781617291807
392 pages, $49.99
February 2017

### Microservices Patterns
*With examples in Java*

by Chris Richardson

ISBN: 9781617294549
520 pages, $49.99
October 2018

### Kafka Streams in Action
*Real-time apps and microservices with the*
*Kafka Streams API*

by William P. Bejeck Jr.

ISBN: 9781617294471
280 pages, $44.99
August 2018

### Streaming Data
*Understanding the real-time pipeline*

by Andrew G. Psaltis

ISBN: 9781617292286
216 pages, $49.99
May 2017

*For ordering information go to www.manning.com*